HACKING MATTER

Also by Wil McCarthy

The Wellstone

The Collapsium

Bloom

Murder in the Solid State

The Fall of Sirius

Flies from the Amber

Aggressor Six

HACKING MATTER

Levitating Chairs, Quantum Mirages,

and the *Infinite Weirdness* of

PROGRAMMABLE ATOMS

WIL McCARTHY

A Member of the
Perseus Books Group
New York

This book's prologue first appeared, in slightly different form, as an essay in *NATURE*. Other chapters are derived in part from material first published in *WIRED*, *Analog*, *Science Fiction Weekly*, *Science Fiction Age*, and assorted other magazines and Internet sites.

Published by Basic Books,
A Member of the Perseus Books Group

Hardback first published in 2003 by Basic Books
Paperback first published in 2004 by Basic Books

Designed by Jeff Williams
Set in 10.5-point Garamond MT by the Perseus Books Group

Library of Congress Cataloging-in-Publication Data

McCarthy, Wil.
Hacking matter : levitating chairs, quantum mirages, and the infinite weirdness of programmable atoms / Wil McCarthy.
p. cm.
Includes index.
ISBN-13 978-0-465-04428-3 (hc)
ISBN-13 978-0-465-04429-0 (pbk)
ISBN-10 0-465-04428-X (hc)
ISBN-10 0-465-04429-8 (pbk)
1. Nanotechnology—Forecasting. 2. Quantum electronics—Forecasting.
I. Title.
T174.7 .M38 2002
620'.5'0112--dc21

2002015887

To Michael and Evalyn McCarthy,

for making me curious

We're always looking for new physics; new behavior that has never been seen before. Once we find it, of course, we start to daydream.

— MARC KASTNER

CONTENTS

JULY 4TH, 2100
PROGRAMMABLE MATTER: A RETROSPECTIVE

THE FLICK OF A SWITCH: a wall becomes a window becomes a hologram generator. Any chair becomes a hypercomputer, any rooftop a power or waste treatment plant. We scarcely notice; programmable matter pervades our homes, our workplaces, our vehicles and environments. There isn't a city on Earth—or Mars, for that matter—that isn't clothed in the stuff from head to toe. But though we rarely stop to consider it, the bones of these cities—their streets, their sewers, the hearts of their telecom networks—were laid out during a time when the properties of matter were dictated exclusively by Mother Nature.

Just imagine: if specific mechanical or electrical properties were desired, one first had to hire miners to extract appropriate elements from the Earth, then chemists and metallurgists to mix precise proportions under precise conditions, then artisans to craft the resulting materials into components, and assemble the components into products that could then be transported to the location of desired use. The inconvenience must have been staggering.

In the twenty-second century, of course, any competent designer will simply define the shape and properties she requires—including "unnatural" traits like superreflectance, refraction matching (invisibility), and electromagnetically reinforced atomic bonds—and then distribute the configura- 

tion file to any interested users. But prior to the invention of wellstone—the earliest form of programmable matter—this would have sounded like pure fantasy. With that in mind, we'll look back on the invention upon which, arguably, our entire modern civilization rests.

Consider silicon, whose oxide is the primary component of rock—literally the commonest material on Earth. Humans had been making hammers and axes and millstones out of it for millions of years, but as it happens, silicon is also a semiconductor—a material capable of conducting electrons only within a narrow energy band.

While silicon was invaluable in the development of twentieth-century digital electronics, its "killer app" eventually proved to be as a storage medium for electrons. When layered in particular ways, doped silica can trap conduction electrons in a membrane so thin that, from one face to the other, their behavior as tiny quantum wave packets takes precedence over their behavior as particles. This structure is called a "quantum well." From there, confining the electrons along a second dimension produces a "quantum wire," and finally, with three dimensions, a "quantum dot."

The unique trait of a quantum dot, as opposed to any other electronic component, is that the electrons trapped in it will arrange themselves as though they were part of an atom, even though there's no atomic nucleus for them to surround. Which atom they emulate depends on the number of electrons and the exact geometry of the wells that confine them, and in fact where a normal atom is spherical, such "designer atoms" can be fashioned into cubes or tetrahedrons or any other shape, and filled with vastly more electrons than any real nucleus could support, to produce "atoms" with properties that simply don't occur in nature.

Significantly, the quantum dots needn't be part of the physical structure of the semiconductor; they can be maintained just beneath the surface

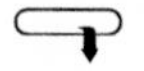

through a careful balancing of electrical charges. In fact, this is the preferred method, since it permits the dots' characteristics to be adjusted without any physical modification of the substrate.

Who "invented" wellstone remains a matter of confusion and debate; similar work was being performed in parallel, in laboratories all over the world. But regardless of where the idea originated, the concept itself is deceptively simple: a lattice of crystalline silicon, superfine threads much thinner than a human hair, crisscrossing to form a translucent structure with roughly the density of polyethylene. Wellstone behaves fundamentally as a semiconductor, except that with the application of electrical currents, its structure can be filled with "atoms" of any desired species, producing a virtual substance with the mass of diffuse silicon, but with the chemical, physical, and electrical properties of some new, hybrid material.

Wellstone iron, for example, is weaker than its natural counterpart, less conductive and ferromagnetic, basically less iron-like, and if you bash it over and over with a golf club it will gradually lose any resemblance to iron, reverting instead to shattered silicon and empty space. On the other hand, it's feather-light, wholly rustproof, and changeable at the flick of a bit into zinc, rubidium, or even otherwise-impossible substances like impervium, the toughest superreflector known.

Of the changes wrought by programmable matter in the past one hundred years, not all have been universally welcomed. In the grand Promethean tradition, wellstone places the power of creation and destruction squarely in human hands. Many have argued that far from making us strong, this power fosters a quiet corruption of spirit. Still, the fable of the three little pigs holds true: not even the Luddites among us build their houses of straw or sticks when impervium is a free download.

ACKNOWLEDGMENTS

THIS BOOK IS THE END PRODUCT of a long evolutionary chain of stories and articles, whose existence and worldly reception owe a great deal to their editors. In roughly chronological order, I'd like to thank Scott Edelman, Chris Schluep, Simon Spanton, Henry Gee, David A. Truesdale, Anne Lesley Groell, Stanley Schmidt, Martha Baer, Chris Anderson, and Bill Frucht.

The basic ideas came about after I'd digested *The Quantum Dot* by Richard Turton and a related article by Ivars Peterson, and discussed them at considerable length with the esteemed doctors Gary E. Snyder and Richard M. Powers. All of the above are responsible for seriously diverting the course of my life. I'd also like to thank Kenneth M. Edwards of the Air Force Research Laboratory for becoming interested at exactly the right time.

For peer review and commentary, I owe a lot to Geoffrey A. Landis, Robert A. Metzger, and Michael P. McCarthy, and for general support I'm grateful to Cathy McCarthy, Nancy Snyder, Shawna McCarthy, David Brin, and the Edge club. I'll also extend a special note of thanks to Vernor Vinge, who encouraged me to pursue the science in a science-fictional idea.

And of course none of these ideas would exist at all without the scientists doing the research. The ones I've spoken with have been uniformly friendly and enthusiastic, and I've tried to do them justice here, with all due apologies to the dozens of other scientists whose fundamental contributions I've left out or overlooked. Too often, the enormous assistance of grad students and postdocs go unrecognized, so I'll also extend blanket thanks to Andrei Kogan, Myrna Vitasovic, Nikolai Zhitenev, Gleb Finkelstein, Stuart Tessmer, and all the rest of you out there.

Any errors in this book should be blamed on entropy, which always increases no matter how hard we try.

Clarke's Law and the Need for Magic

O Nature, and O soul of man! how far beyond all utterance are your linked analogies! Not the smallest atom stirs or lives on matter, but has its cunning duplicate in mind.

—Herman Melville, *Moby Dick* (1851)

THE HARDEST THING YOU CAN ASK THEM is how old they are. The question seems to rock them back, to give them pause. "I guess I'm 38," one of them tells me uncertainly. "I must be 54," another answers, after even longer deliberation. It's not that these men are slow, it's that they're physicists. And they're involved in a research area as promising as it is new and strange, so if they seem a little distracted, well, *c'est la vie.* Despite a cautious modesty so deeply ingrained that it might well be genetic, they also project an air of barely contained excitement. They're building a magical future, and they know it.

Through the entirety of human history, from the moment the first stone was picked up and hurled at an attacking predator, our lives have been shaped and focused and empowered by our technology. Nature would have us naked and unprotected, scrabbling in the dirt for sustenance; we prefer to be clothed and warm, well nourished, and equipped with a variety of tools to shape and interact with the environment around us. Initially these tools were found objects: sticks and stones. Later, we began to shape them for specific purposes, and then to connect them in intricate ways. We progressed from tools—static pieces of specialized matter—to machines, which are tools that can change their shape, and convert energy from one form to another. Matter that works, so you don't have to. Soon, we were experimenting with abacuses, and with animated models of the heavens known as "orreries." These led directly to mechanical calculating machines, and eventually to designs for general-purpose computers—matter that thinks. This idea no longer shocks us—we've lived with computers for too long—but there is nothing natural about it.

Technology is literally the study of technique, but by the twentieth century it had become possible to study technology itself—the changes and directions and underlying motivations of the invented world, and the possibilities that might soon arise. The literature of science fiction took note of these observations, and, indeed, in his 1962 collection *Profiles of the Future*, writer and visionary Sir Arthur C. Clarke formalized three "laws" of technological development:

- First Law: "When a distinguished but elderly scientist states that something is possible, he is almost certainly right. When he states that something is impossible, he is very probably wrong."
- Second Law: "The only way of discovering the limits of the possible is to venture a little way past them into the impossible."
- Third Law: "Any sufficiently advanced technology is indistinguishable from magic."

The first two laws are largely forgotten, and the third, commonly known as "Clarke's Law," was actually stated less succinctly in the 1940s, based on a similar comment made by the alchemist Roger Bacon some 700 years prior, when he wrote of crude eyeglasses and telescopes and microscopes and described them as a "natural magic." What Bacon observed, and Clarke formalized, is that the ultimate aim of technology is simple wish fulfillment.

"Magic" has been technology's partner from the very beginning—a similar attempt to grasp and shape the forces of the world. Any anthropologist will tell you that magic is a rational belief based on sound principles of analogy and empiricism. Unfortunately, it has been far less successful than its partner. We yearn for it, write poems about it, but find no hard evidence for it in our world. The magic we examine turns out to be coincidence or natural processes, or outright trickery. But here is the corollary of Clarke's Law: that trickery is also a technology, and one that fulfills a definite human need. We use the levers and pulleys of technology to shape our world, but what we really want is a world that obeys our spoken commands and reconfigures itself to our unvoiced wishes. What we really want—what we've always wanted—is magic.

The future is where these two notions converge. If matter can work and think, can it also be made to *obey*, at some fundamental, near-magical level? The answer I will give here is not a simple yes or no, but a survey of a class of electronic components called "quantum dots" and their possible application to the fields of computing and materials science.

This is not a book about "nanotechnology" in any of its popular incarnations. Nor will I spend much time discussing the nearer-term technology of microelectromechanical systems, or MEMS, which has already found its way into some applications. The future almost certainly holds myriad uses for both of these, but by the time they find their way into the real world, they may wind up looking less magical than a humble television screen, which after all can change its appearance instantly and completely. But there may be a truly programmable substance in our future that is capable of changing its apparent physical and chemical properties as easily as a TV screen changes color. Call it programmable matter.

A Matter of Scale

Before we begin, it's helpful to clarify the issue of scale. Virtually everyone is familiar with the millimeter (mm), a unit of length equal to one tenth of a centimeter or 0.03937 inches. This is the smallest of the everyday units that nonscientists use in normal life. People in the medical and electronics professions may be almost as familiar with the much smaller micron or micrometer (μm), which represents one thousandth of a millimeter—the primary unit for measuring microscopic things, whether living or non-. Only a handful of professionals (mainly chemists and physicists) are interested in the nanometer (nm), which is one thousandth of a micron or one millionth of a millimeter. This is the scale of molecules, and it is generally invisible to us even with optical microscopes. There are still smaller units, such as the Angstrom (0.1 nm), picometer (0.001 nm), and Planck length (1.610^{-26} nm), but for the purposes of this book, these are awkward and will be avoided. Similarly, since the objects and devices we'll be considering are mainly microscopic, the millimeter is a bloated unit for anything other than occasional reference. For the next seven chapters, we'll be dealing heavily in microns and especially nanometers, so Table 1.1 is provided to show how familiar objects stack up against these measurements. If you get confused later on in the book, checking back here may help.

In the microscopic realm, scale is of critical importance. On the macroscale—the familiar world of meters and millimeters and kilometers—the laws of physics are essentially the same regardless of how big or small an object is. Gravity and electromagnetism have the same effect on stars and planets as they do on pebbles and sand grains. This rule holds true in the upper reaches of the microscopic realm as well: red blood cells (about 10 μm across) can be modeled quite well with the equations of classical dynamics and fluid mechanics.

At the nanoscale, where we find very tiny, very simple objects like the water molecule (about 0.3 nm across at its widest), these rules barely apply at all. Instead, the behavior of particles is governed by quantum

TABLE 1.1 THE SIZES OF THINGS

Item	Size		
Smallest Ant	2 mm	2,000 μm	2,000,000 nm
Largest Protozoan	0.75 mm	750 μm	750,000 nm
Dust Mite	0.25 mm	250 μm	250,000 nm
Human Hair (Diam.)	0.1 mm	100 μm	100,000 nm
Talcum Grain	0.01 mm	10 μm	10,000 nm
Red Blood Cell	0.008 mm	8 μm	8,000 nm
E. coli Bacterium	0.001 mm	1 μm	**1,000 nm**
Smallest Bacterium	0.0002 mm	0.2 μm	**200 nm**
Influenza Virus	0.0001 mm	0.1 μm	**100 nm**
Cellular Membrane	0.00001 mm	0.01 μm	**10 nm**
C_{60} "Buckyball"	1.0×10^{-6} mm	0.001 μm	**1 nm**
Francium Atom	5.0×10^{-7} mm	0.0005 μm	0.5 nm
Oxygen Atom	1.3×10^{-7} mm	0.00013 μm	0.13 nm
Hydrogen Atom	6.0×10^{-8} mm	0.00006 μm	0.06 nm

THE MESOSCALE

mechanics, that elusive and slippery physics pioneered in the time of Einstein. Quantum mechanics is almost completely counterintuitive; your "gut feel" about how a particle should behave is virtually useless for predicting what it will actually do. This is because on the nanoscale, what we call "particles" are really "probability waves"—regions where a particle-like phenomenon is more or less likely to occur. Probability waves can do "impossible" things like leaping across an impenetrable barrier, or existing in many places at the same time, or apparently predicting the future, or being influenced by distant events much faster than the speed of light should allow.

But this intuitive mess is at least orderly in a mathematical sense, and is well described by the "quantum field theory" of the early and middle twentieth century. Small molecules possess a high degree of symmetry and a relatively small number of constituent particles (or waves). As a result, their behavior under various circumstances can be predicted with great accuracy, even though it makes no apparent sense to us as human beings.

So the microscale of the red blood cell is a very different place from the nanoscale of the water molecule. The mathematics that describe them are completely different. But these two scales are separated by three orders of magnitude (i.e., by a factor of 1,000) in size, and between them lies a mysterious realm called the mesoscale (from the Greek "mesos," or middle), where neither set of theories is accurate. The equations of quantum field theory become exponentially more complicated as the number and size of particles increase–especially because random voids and impurities creep in, disrupting the quantum waveforms in unpredictable ways. So while quantum theory is highly accurate, its predictions tend to be almost worthless on the mesoscale.

Similarly, the classical "laws of physics" are really just statistical observations—the averaged behavior of large groups of atoms. But these averages, like any statistics, lose their validity as the sample size decreases. There is no "average particle," just as there's no "average human being" or "average hockey game." Objects much smaller than a micron in size start to behave in some very non-Newtonian ways, as the nonaverage behavior of individual particles increasingly stands out.

So if mesoscale physics are neither classical nor quantum, we clearly need some entirely new set of theories to describe them. This will be an important frontier for physics and chemistry as the twenty-first century unfolds. Meanwhile, our shrinking electronics technology is creeping down into the mesoscale whether we're ready or not, and our increasingly large and sophisticated designer molecules are unfortunately creeping *up* into the same realm from the other direction. Experimentation on the mesoscale—to give us at least some marginal information about where we're headed—has been a hotbed of activity since the late 1980s.

The study of mesoscale effects is an important aspect of "condensed matter physics," which *Britannica* defines as "the study of the thermal, elastic, electrical, magnetic, and optical properties of solid and liquid substances." It is here in this scientific hinterland that we find Drs. Marc Kastner, Moungi Bawendi, Charles Marcus, and Raymond Ashoori. Their specialty: mesoscopic semiconductor structures with bizarre new properties.

Some of their early discoveries are quite astonishing.

2

Standing Waves

> Already we know the varieties of atoms; we are beginning to know the forces that bind them together; soon we shall be doing this in a way to suit our own purposes. The result—not so very distant—will probably be the passing of the age of metals. . . . Instead we should have a world of fabric materials, light and elastic, strong only for the purposes for which they are being used.
>
> —John Bernal, "The World, the Flesh, and the Devil" (speech, 1929)

BOSTON IS A CHALLENGING CITY to get around in even if you have a good map, a good sense of direction, and a mile-wide destination right on the banks of the Charles River. One wrong turn, and you're lost among narrow one-way streets that are hundreds of years older than the automobile. From hilltops and between buildings you may glimpse the bridges that lead across into Cambridge, though never via the road you're actually on. You may even see MIT itself—it's an imposing collection of cement and sandstone edifices, giant columns, and very tall buildings for a university. Of course, there are five other college campuses in the immediate area, plus hospitals and museums and government centers to confuse the unwary. Also identical parking garages that will happily charge you $20 for an afternoon.

But rest assured, you'll get there. MIT is as ugly as it is hard to reach, and it sticks up out of the city like a bouquet of sore thumbs. Universities are quirky by nature, but here quirkiness seems to be a point of pride; the scale of the place, while impressive, dictates that it can't really be imaged or photographed except possibly from the air. The buildings are too large, too close together, too close to other things to fit in a camera lens, and anyway every square and courtyard boasts a big, ugly, nonrepresentational sculpture that seems calculated to repel photographs. Most quirkily, what appear on the map as separate buildings are in fact all connected like a giant Victorian shopping mall, through a gloomy central passage known as the "Infinite Corridor." The occasional south-facing window, looking out across the river at the towers of Boston, half a mile and a few dozen IQ points away, provides the only real connection with the world as most people know it. By American standards this place is *old*—everywhere you find doorknobs worn smooth, staircases bowed like grain chutes, stone benches so old they sag in the middle, flowing over centuries like bench-shaped sculptures of molasses.

The decor is schizophrenic as well: here a row of old classrooms, there a carved monument to the university's war dead. The bulletin boards are adorned with notices of various kinds: some political, some artistic, some related purely to student life. Many of them are ads from and for extremely specialized technical journals, conferences, and job postings. Now we pass through a set of double doors, up a few flights of well-hidden stairs, and into a red-painted hallway lined with pressure tanks and cryogenic dewars and other heavy and vaguely dangerous-looking equipment. Suddenly there are a lot of warning signs:

DANGER: Laser
WARNING: High Magnetic Field
WARNING: Potential Asphyxiant. May cause severe frostbite.
Wear Eye Protection

And my personal favorite,

CAUTION: Designated area for use of particularly hazardous chemicals. May include select carcinogens, reproductive toxins, and substances with a high degree of acute or chronic toxicity. Authorized personnel only.

Interestingly, that notice is dated 1990, and curls up noticeably at its yellowed corners.

We are, to put it mildly, not in Kansas anymore. We've entered the Center for Materials Science and Engineering, a fey landscape whose residents would be appalled to describe it as "magical." The drinking fountains don't work, for one thing. Nonetheless, the air virtually crackles with subdued excitement. Passersby hurry along, animated with arcane knowledge. An enormous amount of science goes on in this building alone, and even a cursory inspection of the jokes and comics, the charts and viewgraphs and magazine articles taped to the walls will let you know, sure enough, that the future is being invented here. Where else? This is MIT, dude—the *present* was invented here thirty years ago.

Marc A. Kastner, Ph.D., is the head of the university's physics department—which is kind of like being the head of the "Boats and Sailors" department of the U.S. Navy. He's a short, wavy-haired man, with a mustache and glasses and a perfectly innocent demeanor. He's old enough to be the father of most of his students, but still seems rather young and rather mild to be in such a responsible position. It's difficult to imagine him chewing anyone out. He's been here at MIT since 1973, so people may simply be afraid to refuse his bidding. He seems like a chap who knows where the bodies are buried.

"I started working with nanostructures around 1978," he tells me, almost as soon as we've sat down. He is full of words, and they bubble over at the slightest provocation. "They raised such interesting physics questions. We were trying to make one-dimensional FETs [a type of transistor], and discovered quantum dots by accident. Our team was the first to add a gate to a quantum dot instead of just two electrodes. We made the first semiconductor single-electron transistors as well—previously, they'd always been made of metal."

His words could not be described as boastful—in fact, Kastner barely seems to appear as a character in his own story. It's the work itself that possesses him. Nor is the discussion aimed above my head; Kastner is, among other things, a teacher. He knows how to simplify and clarify a complex subject. But we've corresponded before this visit, so he has a pretty good idea how much I know, and how much information he can safely dump on me without provoking an allergic reaction. I open my notebook and begin scribbling.

Solid-state physics, more properly known today as condensed matter physics, is a rich and detailed subject that is mostly outside the scope of this book. Still, to understand Kastner's narrative and its implications for the field of materials science, certain narrow slices of physics should first be understood. If you're already familiar with these, then the next few pages can serve as a quick refresher. If not, then treat it as a compressed and somewhat oversimplified introductory lesson, requiring only a high-school familiarity with atoms, electrons, and the flow of electricity. Either way, you'll be through it quickly and (I hope) painlessly, and on to the juicier details of programmable matter.

Semiconductors and Quantum Wells

Most materials are either conductors, which permit the free flow of electrons, or insulators, which resist it. Semiconductors are insulators that are capable of conducting electrons within a certain narrow energy band—a useful trick that makes integrated circuits and other electronics possible. The most familiar semiconductor is silicon, which is used to make the vast majority of microchips found in today's consumer and industrial electronics. Silicon is literally dirt cheap. Its native oxide, SiO_2, is the main component of sand and rocks. When melted, purified, and hardened into sheets, silicon dioxide also serves as one of our favorite insulators and building materials: glass. Unlike most other semiconductors, silicon is also nontoxic.

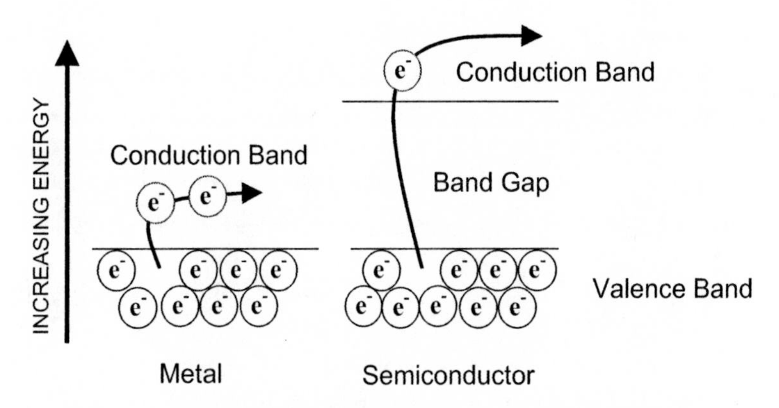

FIGURE 2.1 ENERGY LEVELS OF A METAL AND A SEMICONDUCTOR

In a metal, many electrons rest naturally in the conduction band and can be pushed to neighboring atoms with only a tiny addition of thermal or electrical energy. In a semiconductor, enough energy must first be added to excite the electron out of the valence band and across the band gap. Thus, semiconductors require much higher voltages and temperatures in order to conduct electricity.

The electrical properties of a semiconductor like silicon are of course fixed by the laws of physics. Atoms hold electrons in "shells," which increase in size, capacity, and potential energy the further they are from the nucleus. You can think of shells as layers in which electrons can exist. "Valence" electrons are found in full (or nearly full) shells, where there are few empty spaces for electrons to move through. These electrons tend to stay at home, so their levels exhibit a large electrical resistance, and do not permit electricity to flow. "Conduction" electrons are found in shells that are more than half-empty. These shells have lots of open space, so electrons are free to travel through them. Conduction electrons thus move easily from one atom to another. Between these layers is a "band gap" of forbidden energies. Here, there are no electrons at all, ever. (See Figure 2.1.)

Electrons below the band gap of a semiconductor behave as though they were in an insulator, with nearly infinite electrical resistance, while electrons above the band gap behave as though they were in a conductor. They flow fairly easily, and can be used to move energy and information around.

The difference between a metal and an insulator is that the outermost electron shell of a metal is more than half-empty. It has lots of conduction electrons. An insulating material, such as sulfur, has an outer shell that is almost completely filled. All its electrons are valence electrons—homebodies that don't like to travel. Semiconductors are the fence-sitters. With outer shells that are approximately half-filled, they carry valence electrons that, with the input of energy, can jump to a higher level where they find lots of open space to travel through.

The amount of energy needed to trigger this jump is a property unique to each semiconductor. Interestingly, though, it can be adjusted through a process known as "doping," in which very small and very precise amounts of another material are scattered throughout the semiconductor's crystal lattice.

A material like silicon, when doped with electron "donor" atoms such as phosphorus, becomes an "N" or negative-type semiconductor, which contains one excess electron for every atom of dopant. Often, this doping is controlled almost to the level of individual atoms, and typically about one dopant atom is added per million atoms of substrate. It doesn't sound like much, but this tiny impurity can wreak large changes in the semiconductor's behavior such that, for example, room-temperature electrons have a good chance of jumping up into the conduction band when a voltage is applied. Since they have fewer free electrons than metals do, N-type semiconductors do not conduct electricity as well as metals. But they do conduct it more easily than unaltered semiconductors.

Doping with electron "borrower" atoms like aluminum produces a "P" or positive material, which conducts "holes," or spaces where an electron isn't. Electron holes can be manipulated and moved around as though they were positively charged particles. The analogy is that little puzzle where you slide the squares around to unscramble a picture or a sequence of numbers—you rearrange the puzzle by moving the hole where you want it. Anyway, with "P"-type silicon you get one extra hole per atom of dopant, meaning that a small, precise number of excess electrons can be absorbed by the material, so their free flow is sharply inhibited.

This may sound rather abstract, but it's a trillion-dollar factoid: a "P" layer placed next to an "N" layer creates a structure known as a "P-N junction," which is a kind of electrical valve or gate that permits electrons to flow easily in one direction but not the other. This effect is critical in electronic components such as diodes, LEDs, rectifiers, and transistors. In fact, the latter half of the twentieth century was built almost entirely on P-N junctions; without them, we would not have the compact computers and communication devices that made all the other advances possible.

Since the late 1980s, another application for P-N junctions has been discovered that may, in the end, prove even more revolutionary. When an N layer is sandwiched between two Ps, a kind of "trap" is created that attracts electrons into the middle layer and doesn't let them out. This is a useful trait all by itself, and it leads to a couple of exotic variants on the standard N-P-N transistor. But if the N layer is really thin—about 10 nanometers or 0.000001 millimeters or 50 atoms high—something weird starts to happen: the size of the trap approaches a quantum-mechanical limit, the de Broglie wavelength of a room-temperature electron.

The uneasy science of quantum mechanics, developed in the first half of the twentieth century, tells us that subatomic particles like the electron aren't particles at all but probability waves, regions of space where a particle-like phenomenon may or may not be found. At larger scales—even at the level of microns—the difference is moot, but on the atomic level it becomes critical. Waves travel and interact in ways very different from particles; it's the difference between a feather pillow and an iron bead.

What happens to a P-N-P junction on this scale is critical and fascinating: along the vertical axis of the trap, the excess electrons no longer

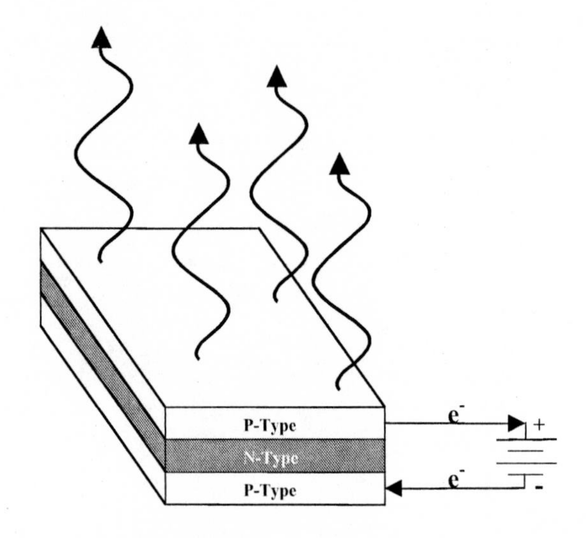

FIGURE 2.2 P-N-P JUNCTION

When a P-N-P junction is thin enough to force wave-like behavior along its vertical dimension, it becomes a "quantum well" that traps electrons in the N layer. At the upper P-N interface, it also brings together large numbers of electrons and "holes" at very precise energies, producing photons at a characteristic wavelength.

have room to move and propagate in the Newtonian way. Their positions and velocities take on an uncertain, probabilistic nature. They become waves rather than particles. (See Figure 2.2.)

Such devices, known as "quantum wells," are easy and cheap to produce. And when voltages are placed across them, they bring large numbers of electrons and electron holes together at fixed energies, and thus have the interesting property of producing photons of very precise wavelength. This means they can be used to make laser beams, including "surface emitting" lasers that can be fashioned directly onto the surface of a microchip. Quantum wells find practical use in optical computers, fiber-optic networks, and those cute little $7 laser pointers you can buy for your keychain.

These wells can be, and for a variety of reasons often are, made from semiconductors other than silicon. (Silicon lasers would be extremely inefficient compared to their gallium arsenide or GaAs equivalents.) Still, the principles are the same either way, and for the sake of simplicity I'll continue to use silicon as a discussion example, with the understanding that more unusual and complicated materials may in fact be the norm for these applications. The important thing to focus on is the electrons themselves.

Now let's talk dimensions: for our purposes, a quantum well has too many of them. It confines electrons in a two-dimensional layer, like the meat inside a sandwich. But if the meat and the top bread layer are sliced away on two sides, leaving a narrow stripe of P-N sandwich on top of a sheet of P bread, the electrons take on wave-like behavior along an additional axis. This structure, called a "quantum wire," is used to produce very intense laser beams that can be switched on and off much more rapidly than quantum well lasers can—up to 40 gigahertz (GHz), or 40 billion times per second, and soon perhaps up to 160 GHz. (See Figure 2.3.) Quantum wires can also be used as optical fibers, as precision waveguides to steer high-frequency electromagnetic signals around in a circuit, and of course as actual wires.

But quantum wires, like quantum wells, are of interest to us here only as a stepping stone. They lead us to a final step: etching away the sides of the stripe to leave a tiny square of meat and bread atop the lower slice, to produce a "quantum dot" that confines the electrons in all three dimensions. (See Figure 2.4.) Unable to flow, unable to move as particles or even to hold a well-defined position, the trapped electrons must instead behave as de Broglie standing waves, or probability density functions, or strangely shaped clouds of diffuse electric charge. "Strangely shaped" because, even as waves, the negatively charged electrons will repel each other and attempt to get as far apart as their energies and geometries permit.

If this sounds familiar, it's because there's another, more ordinary place where electrons behave this way: in atoms. Electrons which are

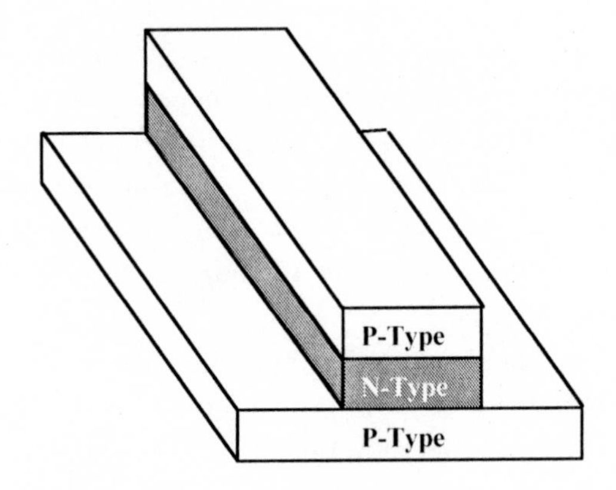

FIGURE 2.3 QUANTUM WIRE

A quantum wire confines wave-like electrons in two dimensions but allows them to propagate along the third (long) axis in a particle-like manner.

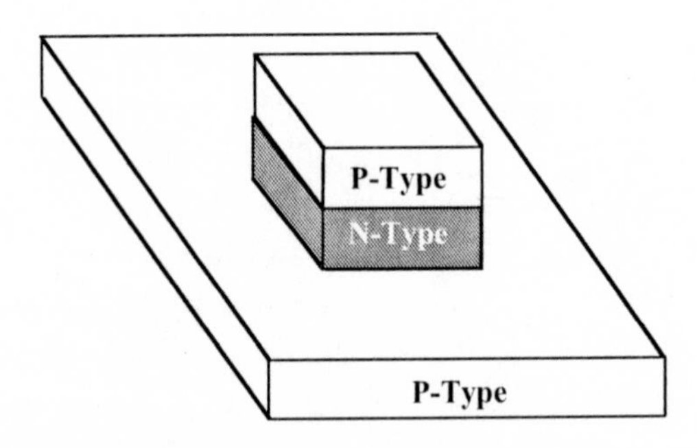

FIGURE 2.4 QUANTUM DOT

A quantum dot confines electrons in all three dimensions, forcing them to behave as standing waves. Their structure thus resembles the electron clouds or "orbitals" of an atom.

part of an atom, will arrange themselves into "orbitals," which constrain and define their positions around the positively charged nucleus. These orbitals, and the electrons that partially or completely fill them, are what determine the physical and chemical properties of an atom—that is, how it is affected by electric and magnetic fields, and also what other sorts of atoms it can react with, and how strongly.

This point bears repeating: the electrons trapped in a quantum dot will arrange themselves as though they were part of an atom, even though there's no atomic nucleus for them to surround. Which atom they resemble depends on the number of excess electrons trapped inside the dot. Amazing, right? If you're not amazed, go back and read the last four paragraphs again. I'll wait.

Ready? Now we'll take it a step further: quantum dots needn't be formed by etching blocks out of a quantum well. Instead, the electrons can be confined electrostatically, by electrodes whose voltage can be varied on demand, like a miniature electric fence or corral. In fact, this is the preferred method, since it permits the dots' characteristics to be adjusted without any physical modification of the underlying material. We can pump electrons in and out simply by varying the voltage on the fence. (See Figure 2.5.)

This is not a science-fictional device but a routine piece of experimental hardware, in daily use in laboratories throughout the world. It was first patented in 1999, by Toshiro Futatsugi of Japan's Fujitsu Corporation, although the technology itself had been under investigation for nearly a decade prior to that. The design for a "single-electron transistor," or SET, had been laid down by the Russian scientist K. K. Likharev in 1984 and implemented by Gerald Dolan and Theodore Fulton at Bell Laboratories in 1987. It was a device made entirely of metal.

This is where Marc Kastner comes in; while he didn't invent this particular device, he did develop a great deal of the late–80s physics and technology that led up to it. Like many such endeavors, it began with a lunchtime challenge from a colleague: any wire can act as an insulator if

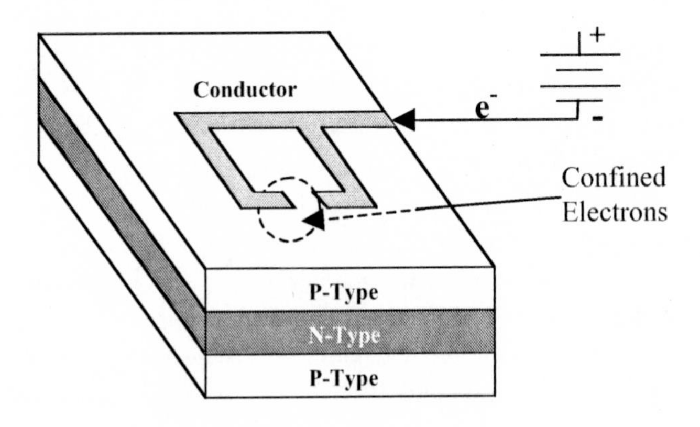

FIGURE 2.5 THE PROGRAMMABLE ATOM

An electrostatic quantum dot uses a voltage to charge a fence-shaped electrode, which confines electrons in the P-N-P junction beneath it. These electrons form an atom-like structure whose properties change as the voltage on the fence is varied.

it's long enough, and the fatter the wire, the longer it needs to be for this to happen. Kastner, working in conjunction with MIT's Electrical Engineering department, was interested in testing this principle with really thin wires, and he was especially interested in making wires with variable resistance. So Kastner and a student named John Scott-Thomas began mucking around with extremely narrow transistors, at which point they bumped into Likharev's mysterious phenomenon known as "quantization of electrical conductance." This led directly to their semi-accidental creation of the first semiconductor SET in 1989, and to the more deliberate invention (in cooperation with the IBM Corporation) of better SET designs shortly thereafter. These perplexing devices later turned out to be one of the many ways to form a quantum dot, which discovery led to Fujitsu's patent.

Kastner is rueful on the subject: "Since the devices were made at IBM, they had the right to patent them but were not interested. I guess

I might have pushed to let us [MIT] get the patent, but I was more interested in publishing the physics."

Because it can be adjusted to resemble any atom on the periodic table, this type of nanostructure is called an "artificial atom"—a term coined by Kastner in 1993 and subsequently taken up widely throughout the semiconductor industry. Other terminology reflects the preoccupations of different branches of research: microelectronics folks may refer to a "single-electron transistor" or "controlled potential barrier," whereas quantum physicists may speak of a "Coulomb island" or "zero-dimensional electron gas" and chemists may speak of a "colloidal nanoparticle" or "semiconductor nanocrystal." All of these terms are, at various times, used interchangeably with "quantum dot," and they refer more or less to the same thing: a trap that confines electrons in all three dimensions.

For the purposes of this book I will use "artificial atom" to refer to the pattern of confined electrons (or other charged particles, as we'll see later on), and "quantum dot" to refer to the physical structure or device that generates this pattern. The distinction is subtle, but clarifying—analogous to the difference between a movie theater and the movie it's currently showing. Other terms, like "single-electron transistor," really reflect *applications* for quantum dot devices rather than the devices themselves, and so will be used only in very limited context.

Building Atoms

Where do quantum dots come from? The poetic answer is that they arise from the sweat and dreams of people like Kastner and the dozens of eager grads and undergrads and postdocs and visiting fellows who work for and with them. There are perhaps forty labs worldwide engaged in this research, and the atmosphere among them is laid back and clubby. This is basic research, geared toward discovery rather than near-term commercial payoff. Collaborations and data sharing are the norm. It's also an extremely young and rapidly advancing field.

"The papers are all recent," Kastner says, gesturing with his hands as if to indicate, somehow, a really short period of time. "The citations are all recent. The papers are on the web prior to formal peer review and publication, because the researchers are so anxious to share results."

The more prosaic answer is that your typical artificial atom comes from the same sort of semiconductor laboratory that produces exotic computer chips. First a salami-sized crystal of silicon or gallium arsenide or what-have-you is grown in a furnace; then it's passed through a machine that melts it, one thin segment at a time. A wave of melting and resolidification passes along the crystal, and when this is complete, all the impurities have been pushed to one end, like the ash on the tip of a cigar. This tip is lopped off, and the remaining ultrapure crystal is sliced into "wafer blanks" that wholesale for around $25 each. An engineering lab then obtains the blank and places it in a molecular-beam epitaxy machine, which sprays a fine, highly customized vapor of semiconductors and dopants, and is capable of building up layers only a few atoms thick, or a few nanometers, or thicker depending on the exact needs of the customer. A scattering of ions may also be implanted, with a device called a "gun."

The resulting finished wafer is shipped to another lab where the final nanostructures are laid down. A metal (usually gold) is deposited over the wafer's surface using chemical vapor deposition (a technique similar to epitaxy) or the "sputtering" of tiny raindrops of molten metal. The wafer is painted with a "photoresist" (a chemical that reacts to light), and the painted surface is exposed, much like a photograph, by an ultraviolet lamp shining through a "mask" in the shape of the desired circuit traces. Where the mask casts a shadow, the resist is not exposed, and remains soft. Everywhere else, the exposed resist hardens into a protective coating. Next, the wafer is dumped in a chemical bath called a "developer," which etches away the soft coating and the metal beneath it, while the hard coating resists the acid, protecting the circuit traces. Then a second chemical or plasma, called the "stripper," is applied to the surface to dissolve away the hardened resist while leaving the metal traces intact.

Even with heroic measures, the finest traces you can hope to produce this way are about 100 nanometers wide, the size of one wavelength of extreme ultraviolet light (EUV). Where smaller structures are desired—and quantum dots can be very fine structures indeed—an additional technique called "electron-beam lithography" etches away still more of the gold, and perhaps select bits of the upper semiconductor layers, leaving behind wires and other traces as thin as 10 nm. This is about the width of 50 silicon atoms—although in real-world practice, 50–100 nm (250–500 atoms) is more typical. Once all the nanostructures are laid down, the wafer is cut crossways into individual microchips, which are then placed into chip sockets of the familiar insectoid design we see soldered to the circuit boards inside any electronic device. Finally, metallic leads are attached to metal "pads" on the chip, and to the insect legs of the chip carrier, by a "gold bonder" device resembling a large sewing machine. (See Figure 2.6.)

The result is a fully programmable quantum-dot-on-a-chip, or even a loose, widely spaced array of several quantum dots. Applying a voltage to certain pins on the chip carrier will pump electrons into the quantum dot, creating an artificial atom and then walking it up step by step on the periodic table. Since an atom's chemical properties are determined by its electrons, these are all you need to create, in chemical terms, an artificial atom. One electron gets you hydrogen, two gets you helium, and so on. Each dot has its own unique periodic table, though; the size and shape and composition of the device have a huge effect on how its electrons interact. We can easily call up an artificial, six-electron "carbon atom" on the chip, but its structure may or may not resemble that of a natural carbon atom.

For example, most quantum dots today are very nearly two-dimensional—the electrostatic "corral" being much larger than the thickness of the quantum well beneath it. This leads to "pancake elements," with a two-dimensional orbital structure that is much simpler than the three-dimensional one for natural atoms. This in turn leads to a more simplistic periodic table, with elements whimsically named for team members

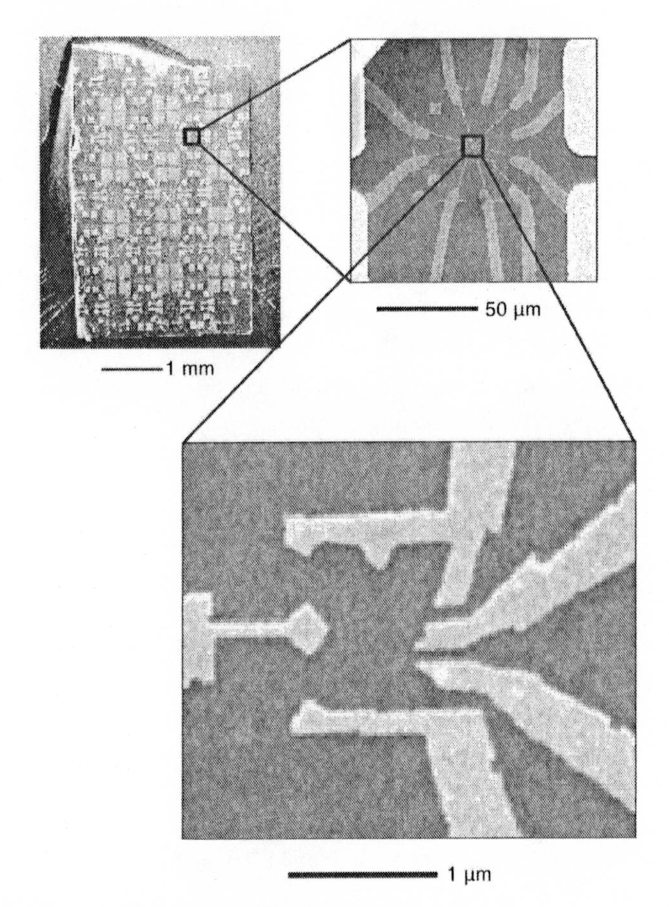

FIGURE 2.6 CHIP MANUFACTURE

The normal techniques of chip manufacture can place dozens of quantum dot devices on a microchip, for control by external hardware. (Image courtesy of Charles Marcus.)

at Delft University and Nippon Telephone and Telegraph. (See Figure 2.7.)

As for the electrical properties of these atoms, the fatherly Kastner explains in patient yet unmistakably excited tones, "The behavior can be completely different from that of the original semiconductor, although

1 Ia							2 Ha
3 Et	4 Au					5 Ko	6 Do
7	8	9 Ho			10 Mi	11 Cr	12 Ja
13	14	15	16	17	18	19	20 Da

FIGURE 2.7 PERIODIC TABLE OF THE PANCAKE ELEMENTS

These two-dimensional atoms, with properties very different from those of natural atoms, are named for researchers at Delft University and NTT. The "Ko" is for Leo Kouwenhoven.

the substrate still has an effect on the final properties." Pump in seventy-nine electrons and what you get is not gold but some related and decidedly improbable material: pseudogold-silicate or pseudogold-cadmium-selenide. And if you force three more electrons into the trap, you can swap atoms of pseudolead in for the pseudogold. Alchemist Roger Bacon, in his monkish, thirteenth-century dreams, could scarcely have asked for more.

Another prediction made by MIT's theorists is that there should be quantum dot materials that behave as insulators when they contain an odd number of electrons, and as conductors (i.e., metals) when they contain an even number. "So far we have not seen this effect," Kastner hastens to explain, "but we're looking hard." He adds, though, that a simple field-effect transistor, or FET, could be considered a "material" with much the same property. The FET is basically an electrically operated switch; when it's open, electrons can't cross it, and the material (or device) is an insulator; when it's closed, electrons have a clear path across it, so it can be considered a conductor.

This is an important point that colors nearly every observation in this book: on the mesoscale and especially on the nanoscale, there is little distinction between a designer *material* and a solid-state electro-optical *device*. In both cases, the aim is to control the properties of matter through careful design, so that the movement of electrons or the application of electromagnetic fields will produce some desired effect. Designer molecules and atomically precise electronic circuits fall precisely into this middle ground, leading once again to a gray area in the standard vocabulary. Therefore, I'll generally use the word "device" to refer to specific nanostructures, such as an individual quantum dot. Large collections of quantum dots, along with the metals and semiconductor substrate from which they're made, will be referred to as "programmable materials" or "programmable matter." This distinction is mushy at best, but it does help to nail down the vocabulary.

In the 1990s, the term "programmable matter" actually enjoyed a limited popularity among cellular automaton enthusiasts. A cellular automaton is a type of computer, or computer program, where calculations are stored in "cells" whose value or output is based on the values in neighboring cells. The most familiar example of this is spreadsheet software such as Lotus 1–2–3 or Microsoft Excel, but it also turns out to be a great way to model things like weather and jet engines. The "programmable matter" label appealed because in cellular simulations of fluid mechanics, the material represented in individual cells could be converted instantaneously from fluid to solid and back again. Walls could be rearranged, or their surface characteristics modified. One fluid could be substituted for another in midflow.

Programmable, yeah, but I submit that this sort of pseudographical mucking around is actually *virtual* programmable matter, whereas our quantum dot chip is the real thing. Precision in language is both beautiful and efficient, so in various speeches and articles I have openly hijacked the term on behalf of the mesoscopic physics community, and offered to thumb-wrestle any cellular automatist who wants it back.

Unusual Computing Properties

Mesoscopic condensed matter physicists do not, in general, like to talk about materials science. It's not really their thing, not quite. In contrast, the marriage of quantum dots with microcomputers is a natural: easy to explain, to fund, to justify. What Kastner and his associates offer up in abundance is cutting-edge computer jargon.

"The most transforming phenomenon of our time," Kastner explains with the look of an enthusiastic but oft-disappointed film critic, "is the invention of the transistor. Moore's first law, which says that the number of transistors on a chip doubles every eighteen months, cannot go on forever. There are fundamental limits to how small transistors can be made."

Through a phenomenon called "quantum tunneling," electrons can in essence teleport from one location to another, across a classically impenetrable energy barrier. In fact, a kind of chemistry takes place even at temperatures near absolute zero, as individual atoms shuffle electrons back and forth in this way. Unfortunately, for a circuit designer, this tunneling is the electronic equivalent of leaky pipes, and the smaller the pipes are, the leakier they get. For transistors much smaller than about 10 nm, the leakage currents are close enough to the operational current that the device is essentially short-circuited, and ceases to perform useful work. This is the dreaded "wall" that chip designers have been anticipating for decades. It isn't upon us yet, but if Moore's Law holds true until 2010, it will be.

Kastner's quiet rant continues: "Probably more important is Moore's second law, which is the observation that the [commercial] facilities to produce the devices double in cost every four years. The generation of facilities built five years ago cost $1 billion each, and 100 of them were built. I don't know how many of the current $2 billion facilities are planned, but it seems clear that investments of this magnitude cannot be maintained. One answer is to somehow give each transistor more functionality, rather than simply increasing the number on a chip."

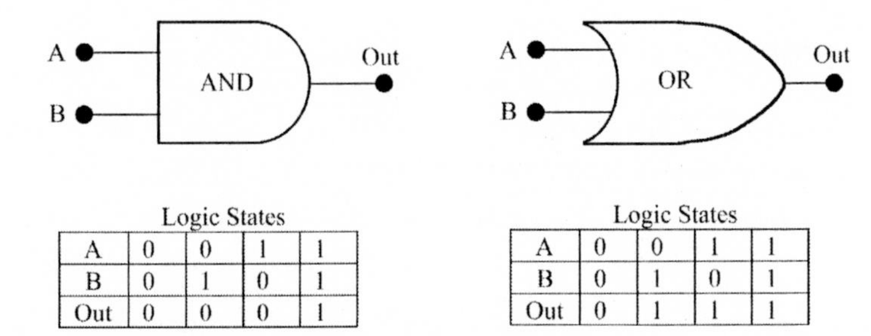

Logic States

A	0	0	1	1
B	0	1	0	1
Out	0	0	0	1

Logic States

A	0	0	1	1
B	0	1	0	1
Out	0	1	1	1

FIGURE 2.8 STATES OF AND AND OR GATES

The AND gate delivers a "true" if both A and B are true. The OR gate delivers a "true" if either A or B is true. Gates like these form the basis of most modern computer circuits.

Now certainly, quantum dots are very, very small. You can fit enormous numbers of them onto a chip. But people familiar with the Boolean logic on which computers are based see another advantage as well. Computer logic is built up of "gates," which compare two binary values (i.e., bits, which hold either a 0 or a 1) and make decisions based on the results. The most familiar of these are the "AND" gate, which outputs a 1 if both of its inputs are 1, and the "OR" gate, which outputs a 1 if *either* of its inputs is 1. (See Figure 2.8.)

Many other types of gates are based on negative logic, such as the "Not-AND" or "NAND" gate, which produces the exact opposite result from the AND gate. Another property often employed is the exclusion of superfluous states. The "eXclusive OR" or "XOR" gate produces a nonzero output only when one input bit is a 1 and the other is a 0. If both inputs are 1 or both inputs are 0, then the gate's output will be a 0. So out of four possible states for the gate, two are the same as for a regular OR gate, and two are reversed. (See Figure 2.9.)

If these logical units seem complex, just imagine the problems of stringing dozens, thousands, or even billions of gates together! Keeping track of such mushrooming complexity is one of the reasons we invented

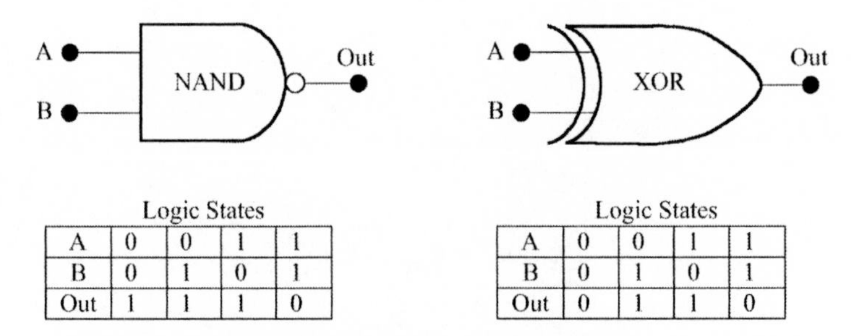

A	0	0	1	1
B	0	1	0	1
Out	1	1	1	0

A	0	0	1	1
B	0	1	0	1
Out	0	1	1	0

FIGURE 2.9 STATES OF NAND AND XOR GATES

The NAND gate delivers a "true" unless both A and B are true. The XOR gate delivers a "true" if either A or B, but not both, are true.

computers in the first place. The point to remember here is that digital computers can be made, with varying efficiency, from a variety of different gate types.

In today's microchips, binary digits are represented with switching devices called metal oxide semiconductor field-effect transistors, or MOSFETs, which are easy to produce with nothing but (nearly) two-dimensional structures of metal, insulative oxides such as SiO_2, and semiconductors. Many other types of switches have been used for computers, including vacuum tubes and mechanical relays, but MOSFETs are cheaper, more reliable, and much easier to miniaturize, so they've largely taken over the world. And because of the way MOSFETs function, we find it easiest to build up computers out of Boolean gates using two to four MOSFETs each.

It turns out, however, that quantum dot transistors behave as eXclusive NOT OR (XNOR) gates—also known as "equivalence gates" because they return a positive if the two inputs are identical, and a zero if they're different. (See Figure 2.10.) This structure would ordinarily require at least eight MOSFETs to assemble, so for some types of calculations, quantum dots are around four times more efficient than

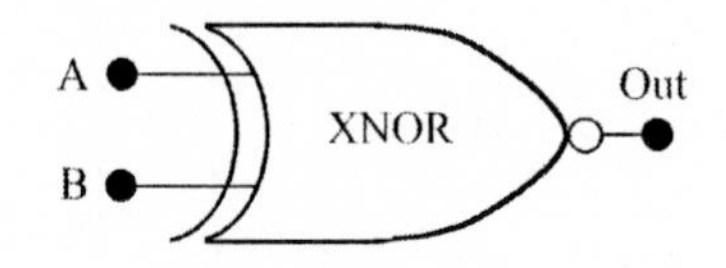

Logic States

A	0	0	1	1
B	0	1	0	1
Out	1	0	0	1

FIGURE 2.10 THE XNOR OR EQUIVALENCE GATE
The XNOR or equivalence gate delivers a "true" if A and B have the same value.

MOSFETs. For certain highly specialized uses such as error correction circuits, they can be as much as twenty-four times more efficient. More important, a circa 2002 MOSFET operating at 1.5 volts requires the passage of about 1,000 electrons in order to change its state from "on" to "off," whereas Kastner's team had built a quantum dot transistor that would change state with the passage of a single electron.

This device led directly to the discovery of the artificial atom as an application for quantum dots. Kastner emphasizes this point philosophically: "As a scientist, it's my goal to understand nature. We actually stumbled onto artificial atoms by accident, while trying to understand something else. We're always looking for new physics, new behavior that has never been seen before. Once we find it, of course, we start to daydream." Variations on this refrain are repeated by nearly every scientist in the field: their fertile utopia is not a set of specific commercial goals but an endless, aimless farting around. Discovery, they insist, happens not on a schedule, but on a whim of nature, for those who happen on some particular day to have the right combination of equipment and curiosity and luck. This of course creates the funding paradox that plagues them endlessly: finding people to bankroll their utopia is difficult when the time and nature of the payoff are unknown.

Anyway, single-electron transistors, or SETs, are effectively little turnstiles, which count the passage of every electron that passes through them. They permit, for the first time in history, the construction of electronic computers that use individual electrons, rather than large bunches of them, to carry information. The energy savings alone are probably worth the cost of retooling, but as a bonus, SET switching times are also much shorter than we're accustomed to, although we'll need new circuit designs to take full advantage of this. So in terms of operations per second—ever the gold standard of computing—quantum dot arrays are probably tens or hundreds of times more powerful than the equivalent MOSFET arrays. And unlike MOSFETs, SETs actually work better the smaller they are. The only problem is manufacturing: making components fine enough and close enough together that the transistor can operate at room temperature. Today's experimental SETs are sometimes a micron or more in size, and they often work only in liquid helium or are so fragile at room temperature that they cease functioning within minutes.

The other problem with SETs is that they don't eliminate the "interconnect problem" of having to physically connect the dots with metallic traces. This need places huge demands on any manufacturing apparatus, whose tolerances would need to be fantastically tight by today's standards. This is certainly possible, but while we're working on it we may find that there are easier cats to skin. "For example," Kastner says, "quantum tunneling means that quantum dots can interact capacitively rather than by current flow through wires." In other words, they can trade electrons across an insulating barrier. Wireless: it's the difference between a radio and a telegraph, or a cell phone and a can on a string.

And when their interactions result from this quantum tunneling of electrons, quantum dots can collectively behave as a form of quantum cellular automaton, or QCA. Today's ubiquitous silicon MOSFET transistors can also be configured this way, of course, and often are, but a lot more of them are needed to make it work. In fact, most of today's cellular automatons are computer programs running on top of operating systems running on top of Basic Input Output Systems (BIOSes) running on top of processors that use MOSFET arrays for memory. When your goal is rapid calculation, this sort of layered architecture is staggeringly inefficient—employed only because it's cheap and quick to set up. But with quantum dot arrays exhibiting QCA behavior inherently, we may eventually see new sorts of computers, capable of solving any problem but by methods completely different from those of today's computers.

Quantum scientists have also shown that an array of SETs passing tunneled electrons around is, mathematically speaking, a form of neural network. This is a type of computer (or sometimes a type of software program) that imitates certain properties of the human brain. In Kastner's words, "it displays associative memory—the way our mind works, as opposed to the way a digital computer works. Only we don't know how to get the information in and out yet, and we don't know how to make it learn." Still, if these problems can be solved, quantum dot "neurons" could display many of the same traits as biological ones, despite being hundreds or thousands of times smaller.

Stranger still, if a phenomenon called "decoherence" can be avoided, the quantum dot can behave as something quite extraordinary: a quantum bit. The study and suppression of decoherence is a major feature of quantum dot research. Why? Because quantum bits are pure freakin' magic. While a normal digital bit can hold only a 1 or a 0, a "quantum bit" (qbit) holds a "superposition of states" that can be treated as a 0 or a 1 or, rather mysteriously, both at the same time. However nutty this may sound, it is solidly factual; qbits were first demonstrated in 1995 by the National Institute of Standards and Technology (NIST) in Boulder, Colorado, and by the California Institute of Technology in Pasadena.

While one qbit can decohere into only two possible states (on or off), five qbits together (the largest number assembled into a quantum computer as of this writing) can collapse into 2^5, or 32, different states. And before the collapse, the qbits, in a very literal sense, can both store and perform computations on all 32 states, a feat that a binary computer would need 32 sets of 5, or 160 bits in all—plus calculating hardware—to match. To put it mildly, qbits provide a certain degree of parallelism, and are somewhat interesting to people focused on parallel computing applications.

Codebreakers are one example: they're constantly searching for numerical "keys" to unlock encrypted messages, but the number of possible values for the key is 2^n, or two raised to the power of the number of bits. To solve a secure 64-bit encryption key for today's Digital Encryption Standard (DES), a Pentium-class digital computer pulling 2 billion calculations per second (2 gigaflop) would require 2^{64} or 1.84×10^{19} operations, or 292.5 years. By contrast, a 64-qbit quantum computer could solve for that same key *in a single operation.* Somewhat interesting, yes.

Since computing power increases exponentially with the number of qbits (versus linearly with the number of digital MOSFET bits), a 64-qbit computer is roughly 18 billion billion times as powerful as a 64-bit binary one. And a 65-qbit computer is twice as powerful as this hypothetical codebreaking dynamo, while a 72-qbit computer is 128 times more powerful still. The raw calculating might of one kilo-qbit (1Kqb) is difficult even to imagine, but of course with quantum dots, building a mega-qbit or even a tera-qbit computer is almost as easy. In a quantum dot world, the concept of "operations per second" as a computing bottleneck simply vanishes. A good desktop metric might be the "EM," equal to the computing power of the entire Earth at the turn of the millennium.

Many other numerical problems are difficult or impossible to solve on present-day computers. They include long-range weather and climate forecasting, the modeling of chaotic systems, and the study of protein folding. Proteins are chain-like molecules that fold up into complex three-dimensional structures, but predicting these shapes from the protein's amino acid sequence has proven surprisingly difficult. The folding appears to be a highly dynamic process, with many intermediate steps before the molecule finally settles into its stable configuration. Present research relies on huge banks of classical supercomputers, and the results are mixed at best. The most significant advances will be in the field of artificial intelligence. Opinions vary widely on how smart or intelligent or conscious a machine can ever be, but as IBM's chess champion Deep Blue demonstrates, even a stupid machine can appear brilliant if it runs fast enough. This is known in industry as "weak AI," but with the power of quantum computing behind it, the results will be anything but weak.

There is of course no right way to design a computer, only better and worse ways for different kinds of operations. In the future, quantum dot computers may well combine all the best features of digital, analog, and quantum processing. The same hardware could in principle be used for any and all of these operations and could switch modes dynamically, depending on the nature of the problem being solved, either globally or else on select tiny regions of a chip. The word "supercomputer" doesn't begin to address the power of such a device, which could model and simulate an entire human brain, as well as perform numerical and quantum-mechanical operations in the tens or hundreds or millions of EM. In other writings I've used the term "hypercomputer" to describe this technology, and in fact I would strongly discourage the use of this term for any lesser device. Otherwise, when the technology actually matures, we'll have run out of cool words to describe it.

The Quantum Laboratory

The artificial atom has also been a real boon to quantum physicists studying the behavior of electrons, because it gives them a new window into the atom. The growing trend in condensed matter physics is for researchers to study man-made objects rather than natural ones like atoms. Because their structure relies not on the electrons' attraction to a positively charged nucleus but, instead, on electrostatic repulsion and the geometry of P-N junctions, artificial atoms are many times larger than natural ones. (See Figure 2.11.) A quantum dot 20 nanometers

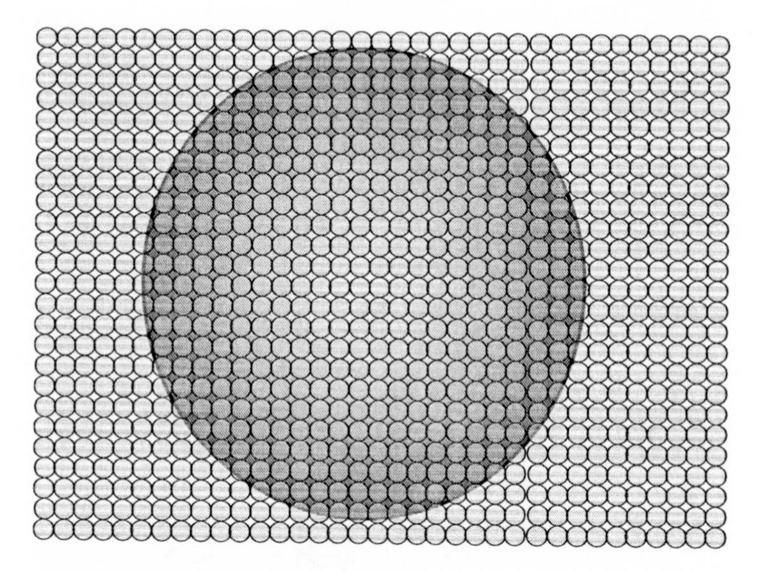

FIGURE 2.11 THE ARTIFICIAL ATOM

The artificial atom (large sphere) does not have a nucleus of its own but can only exist inside an ordered crystal of real semiconductor atoms (small spheres). These real atoms each have a nucleus surrounded by the usual cloud of orbiting valence electrons. The artificial atom is composed of excess (conduction) electrons trapped in the crystal as standing waves.

across would contain an artificial atom roughly fifty times wider than the natural atoms that actually make up the dot. In fact, the artificial atom would physically overlap with these natural atoms. That's an important difference, which in some ways makes it much easier to study. Artificial atoms of this sort are also vastly more flexible than natural ones—there's no need to hunt down particular elements or isotopes in order to perform an experiment. Just alter the voltage on the electrodes until you have precisely the atom you want.

Also, notably, the shapes of natural atoms—spherical, dumbbell, pinwheel, tetrahedral, and so on—arise from the purely spherical symmetry of the nucleus' electric field. Artificial atoms do not share this characteristic. Depending on the shape and voltage of the repulsive electrode fence, they can have square or triangular or any other sort of symmetry. They can be shaped like rods or pancakes. Multipart fences with different voltages on each section can even create atoms that are asymmetrical, or that change their shape or size.

What good are properties like these? It's hard to say at this point, although the gut feeling is one of great importance. One tempting speculation is that quantum dots may soon yield a new breed of high-temperature superconductors. The superconductor is one of those nearly-too-good-to-be-true discoveries that arrive like occasional gifts from God. In a superconductor, electrons travel with zero electrical resistance. Not minimal resistance, not immeasurably small resistance, but actually zero. This trait is incredibly useful for certain applications, such as power transmission and extremely strong electromagnets. But it's also incredibly difficult to exploit, because our present superconductors work only at extremely cold temperatures.

The best ones today—operating at temperatures above 77°K (–196°C or –320°F)—are yttrium-barium-copper oxides. These ceramic materials can be cooled in liquid nitrogen, which is condensed right out of the atmosphere and is cheaper than beer. The older metallic superconductors require much colder temperatures and typically have to be immersed in liquid helium in order to work. Unfortunately, while the operation of these "classical" superconductors is completely understood, the newer high-temperature ones present a theoretical quagmire. Nobody is really sure how they work, or why. There are even some indications that pure carbon can act as a high-temperature superconductor, although there may yet be problems with this theory.

I can't help asking the question: "If this were understood, Doctor, and we knew exactly what material properties would make an optimal superconductor. . . . Right now we're discovering these materials by accident, by trial and error. Would quantum dots allow us to design them intentionally? Could we develop superconductors that work at room temperature, or higher?"

Predicting the future isn't Kastner's job, and for the first time his smile looks a little uneasy. Still, the gleam in his eye is unmistakable. "I think artificial atoms have enormous potential, but the killer application is not yet clear. That would be the dream, yes, wouldn't it?"

3

The Play of Light

The intense atom glows
A moment, then is quenched. . . .
—Percy Bysshe Shelley, "Adonais" (1821)

Self-kindled every atom glows,
And hints the future which it owes.
—Ralph Waldo Emerson, "Nature" (1844)

SO FAR I'VE DESCRIBED two forms of quantum dot: etched blocks of quantum well and electrostatic fences. Many other forms exist, such as the quantum corral: a simple ring of atoms on a flat surface. When appropriately small and appropriately cold, this simplest of nanostructures can trap electrons in a two-dimensional region, producing some of the same quantum effects we've already discussed. Quantum corrals are extremely fragile, though, and thus not good for much in the real world.

One of the most interesting forms of quantum dot was also the first to be invented. It was not part of any electronic apparatus and, in fact, was created by people who had no idea that quantum mechanics existed. They were the glassblowers of Renaissance Italy and Germany.

The first true glass ever made by humans was probably in the translucent beads used by Mesopotamians as jewelry around 2500 B.C. At its crudest, glassmaking simply involves the melting and resolidification of sand, not unlike the kiln firing (a.k.a. vitrification or "turning into glass") of clay pottery, which had already been going on for at least 5,000 years. The art of glassblowing probably dates back to 1450 B.C., when the first known glass bottles were stamped with the glyph of the Egyptian pharaoh Thutmose III. Sand occurs in a wide variety of types, and by fiddling with the grain size and composition, early glassblowers found that they could vary the color, and also the melting point and therefore the physical and optical quality of the glass. Popular additives included ground seashells, ash, and metal oxides as coloring agents. Flat panes of glass, suitable for use in stained-glass windows, first appeared in ancient Rome but were not perfected until the twelfth century. Even then, it proved very difficult to make the sort of clear, colorless, optically flat and bubble-free windowpanes we take for granted today.

Thus the general awe at the advent of "crystal"—a highly doped and highly transparent glass first developed in the 1600s. Just as dopants can alter the electrical properties of a semiconductor, selected impurities can also affect the optical properties of a light-transmitting substance such as glass. European crystal contained significant amounts of lead oxide, which made the glass soft enough to be cut or engraved without cracking, and also, incidentally, dramatically increased its index of refraction. Leaded crystal shoots rainbows off in every direction, in a beautiful way that cut glass does not. This is one reason it remains popular today, even though we know that the lead is toxic and leaches out into solvents like water and alcohol.

Somewhere along the line, at a time and place no one now remembers, one of these Renaissance artisans discovered an additive—cadmium selenide (CdSe)—that produced a strong red color unlike anything that could be achieved with silicate minerals or metal oxides alone. But sometimes, crystals doped with this material displayed colors that were quite different, not red at all. Some of them were even fluorescent. The explanation for this property remained a mystery untii very recently. Since CdSe is a semiconductor with properties and molecular structure similar to silicon, A. I. Ekimov and Alexander Efros of Russia's Yoffe

Institute reasoned in the early 1980s that nanometer-sized particles of this stuff, with extremely high surface-to-volume ratios, could be trapping electrons in interesting ways. This in turn might affect the crystal's response to electromagnetic fields—that is, the absorption, reflection, refraction, and emission of light.

The two men had no way of knowing how true and how profound these observations would turn out to be. They were after something quite specific: improved solar panels. Around the same time, though, one Louis E. Brus, a physical chemist at Bell Laboratories, was investigating the behavior of semiconductor nanoparticles suspended in liquids. He found not only fluorescence—which occurs in a variety of natural materials—but "fluorescence intermittency" or regular flashing, which does not. Flashing is normally a property of technological artifacts—complex ones with power sources and switches and light-emitting elements. To observe this behavior in something as simple as a semiconductor crystal, powered by nothing more than ambient indoor lighting, was, well, unexpected.

Following up on the matter, Brus developed a rather ingenious chemical process for the controlled growth of nanoscopic CdSe crystals in an organic solvent. The size of the resulting particles was remarkably uniform, and it turned out to have dramatic effects on the color of the light they emitted. From a single material, Brus found he could produce any color in the visible spectrum, from deep (almost infra-) reds to screaming (almost ultra-) violets. A number of applications, including light-emitting diodes, occurred immediately to Brus and members of his team. But the issues ran much deeper. The color was a function of electron energies inside the CdSe particles, and the electron energies were a function of particle size. The team was bumping around the edges of a new phenomenon, which would eventually come to be known as "quantum confinement."

In Brus' suspensions, in the photoelectrochemical preparations of Ekimov and Efros, and even in the crystal goblets and decanters of some Renaissance glassmakers, CdSe particles were acting as quantum dots, trapping electrons not with electrostatic repulsion but with simple geometry: glass is an insulator, so electrons can flow inside the semi-conductor particle but aren't able to leave it. If the particle is small enough—say, close to the 10 nm de Broglie wavelength of a room-temperature electron—then an artificial atom will form inside it. The structure and properties of this atom will depend on the number of free electrons in the CdSe particle, which in turn depends on its size and purity. These traits could hardly be uniform or well controlled in Renaissance crystal, since the people making it had only the vaguest idea what they were doing, but in general there'd be a large number of very large dopant atoms in the crystal, which bore little resemblance to the natural atoms of silicon, oxygen, lead, carbon, calcium, cadmium, and selenium of which the thing was actually made.

So the tuning of matter with artificial atoms began in an age when the atom itself was considered a myth.

By 1988, the Bell Labs team included Moungi Bawendi, a young chemist with a freshly minted U. of Chicago Ph.D. Bawendi stayed with Brus until 1990, soaking up knowledge and then finally carting it off to the ivory (well, concrete) towers of MIT, where he became a professor. Bawendi has proven to be a bit of a prizewinner: since his arrival he's garnered a Camille and Henry Dreyfus Foundation New Faculty Award, a National Science Foundation Young Investigator Award, a Nobel Laureate Signature Award for Graduate Education, a Packard Fellowship, and Harvard's own Wilson Award. I don't actually know what these are or what they mean, but they sound impressive, and MIT's web pages trumpet them loudly. They are proud of their guy.

And why not? Bawendi may not be one of the quantum dot's original pioneers, but he came into the field early, and his contributions are fundamental. In a 1993 paper, he outlined an improved process for growing cadmium selenide quantum dots in solution. As of this writing, it's still the favored process, and the subject of considerable commercial and scientific interest, because it also permits the dots to dissolve in water. Not in the sense of unraveling into component molecules the way a sugar crystal might, but in the sense of dispersing evenly throughout the water as whole particles. A "solution" of quantum dots contains roughly equal concentrations wherever you look, whereas a "suspension" of them will be more concentrated toward the bottom, as the particles slowly settle out like grains of sand. This happens because soluble particles surround themselves with water molecules, and so behave as though they were part of the liquid itself. Nonsoluble particles simply bounce around between the H_2O molecules, whose buffeting can't suspend them indefinitely against gravity. Think of a heavy toddler in a playpen filled with plastic balls.

Quantum Dot Corporation

Not surprisingly, many people want to live forever, and are willing to shell out enormous amounts of cash to prolong their lives. People also want to stay healthy and fit, and if anything they'll shell out *more* money to achieve this goal. Medical testing—the studying of the body and its components, and the identifying of minor ailments before they become major ones—is big business. It's Big Government too; given a choice between funding condensed matter physics research and funding an equally pricey cure for cancer, most politicians know the answer without even pausing to think. And the gap is widening, as medicine increasingly shows that it can deliver on its promises. Federal investment in life sciences research was about equal in 1970 to its investment in physics and engineering. Between that time and the year 2000, funding in physics and engineering remained flat, while life sciences increased fourfold, to nearly $20 billion per year. That's a lot of aspirin.

Joel Martin, founder and chairman of Quantum Dot Corporation in Silicon Valley, takes Bawendi's water-soluble dots where the money is: into the realm of the biological sciences. A Valley insider and an MBA'ed venture capitalist, Martin (with serious nudging from Berkeley chemist Paul Alivisatos) saw the commercial potential in Bawendi's invention, and in 1998 he set about acquiring the exclusive license on it from MIT and hustling it as quickly as possible from the laboratory to the marketplace.

Martin's Qdot™ particles glow furiously under any light source, and although at 10–30 nm in size they're much too small to see without an electron microscope, they're bright enough to show up as winking, monochromatic pinpoints in an ordinary optical microscope. When chemical receptors are attached to them, they seek out and illuminate individual molecules, allowing biologists for the first time to observe the chemical processes going on inside a cell, *in vivo*, while they're actually happening. With different types of molecules each tagged with a different color, the cell becomes a kind of city map, alive with twinkling traffic.

"It takes the hubris of entrepreneurs to start a company," Martin opines. "This is easy, this is straightforward. But technology rarely complies with that viewpoint. You don't have the shoulders of giants to stand on here; you have to make all your own mistakes. Knowing a little about the science doesn't tell you anything about the problems of manufacturing—you have to get in there in a very Edisonian way and just try all the combinations. But in doing that, you create a huge pool of intellectual property and know-how, which can be used in the future. I give a world of credit to our technical team, who really bust their tails."

At last count, that technical team was fifty-three people strong. Bawendi, a shareholder sitting on the company's Scientific Advisory Board, is a part of the team as well. So is Martin himself; he's got a Ph.D. of his own, and before his involvement in a string of high-tech startups, he did a stint as a college professor. Reflecting on the direct, focused, business-like manner of his speech, I can't help feeling a twinge of sympathy for the cowed undergraduates his younger self addressed.

This same manner makes him a wonderful interview subject, though; in twenty minutes I've learned more about the business jargon of Silicon Valley than I could possibly have imagined. The company, it turns out, has a total cap of $37 million and a burn rate of $600,000 per month. This is good, though, because it's pre-IPO and has an estimated valuation of over $80 million. Its product line is extremely high margin for such a small company, cutting edge but not bleeding edge, which is why Red Herring has it on the "Ten to Watch" list. There are three VCs on the board, from three different investment firms, and boy are they pleased.

Just when Martin is at his most capitalistic, though, the scientist in him reasserts itself. "I'm simply interested in nanoscale devices. I have ideas outside of life sciences. There's really not enough cross-fertilization, though, to leverage breakthroughs from one field into another. To get a business off the ground, to get a product out fast enough to satisfy venture capital's time horizons . . . that takes a lot of focus."

"What other ideas?" I ask him, though I'm not at all certain he'll answer. This is Silicon Valley, where intellectual property is king, and cellphone chatter is severely frowned upon. The walls have ears.

Martin's smile is sly; he's not going to tip his hand, but neither is he going to let such an interesting question hang there unanswered. "Quantum dots will find application in things you're already familiar with today. Detectors and fiber-optic repeaters. Also paints, coatings, security inks. . . . " He trails off suggestively, then shrugs. "Making things better."

Nanoparticle Films

Another of Moungi Bawendi's contributions came back at Bell Labs, when he hit on the idea of settling his quantum dots out of their wet solution and into solid, dry, organized crystals. Today this is another hot area of research.

The electrostatic and quantum-mechanical interactions between atoms have a curious side effect: on the nanoscale, certain kinds of structures will spontaneously self-assemble and self-organize. Some molecules line up in orderly rows, especially in association with certain other molecules or when placed on certain surfaces. This principle is critical in biology—life probably could not have originated without it—

but it's increasingly important in materials science and nanoelectronics as well. The "design" of some types of nanostructures relies on the organizing power of nature.

This may be less mysterious than it sounds at first; Bawendi himself dismisses the idea with a macroscopic example. "Throw oranges in a crate and they will stack themselves into a minimum-energy lattice. I can do the same thing with soccer balls, cannon balls, anything round. They form a crystal; so what?"

Hmm.

Bawendi's process is fascinating to watch, and simple enough that his students can perform it unsupervised. This is saying a lot, because two of the initial ingredients are "organometallic compounds" of cadmium and selenium that are unbelievably toxic, and plastered with warning labels that prohibit their handling except in a glovebox. Have you seen a glovebox? America's favorite nuclear technician, Homer Simpson, occasionally uses one to handle glowing green substances that look, really, a lot like the vials of quantum dot solution in Bawendi's lab. This is another way of saying that the vials themselves, glowing eerily, look like something out of a cartoon. They really do.

The box is simply a vented chemist's hood—not unlike the fan hood above your stove—enclosed in clear plexiglass. Its interior is accessible only through a pair of heavy black gloves, which project through the plastic and are sealed tightly against it at the edges. Chemicals, tools, containers, and glassware are passed in and out of the box through a little airlock. It's a simple device, but to my eye a rather unsettling one. Still, Bawendi's student, Myrna Vitasovic, seems comfortable enough. It takes her about fifteen minutes to heat and mix the appropriate chemicals in there, and when she's finished, the solution takes on the deep, wine-red color that so impressed the glassblowers of Europe—the color of cadmium selenide molecules. These are fortunately less toxic than their organic precursors, so the little beaker can be removed from the glovebox and set on a tabletop to cool.

By this point, though, interactions between the molecules of the organic solvent are causing CdSe crystals to grow symmetrically, at slow and predictable rates, inside organic molecular jackets. Within about one minute, the particles are 10 nanometers in diameter, and the color of the solution changes abruptly from dark cadmium red to dark blue and then, at 14 nm a few moments later, to a bright, liquid, fluorescent yellow that reminds me of a plastic drafting ruler I once owned.

This is interesting all by itself, because a wavelength of yellow light is 580 nm—over forty times larger than the particles in Vitasovic's beaker. Why this particle size should result in this particular color is not at all obvious, and indeed, Brus and his fellows were mightily perplexed by such phenomena. Bawendi had told me: "These optical properties were studied and abandoned ten years ago. Today the physics is better understood."

Basically, light emission from bulk semiconductors (e.g., in an LED) results from electrical or optical excitation of electrons in the semiconductor, creating electron-hole pairs that, when they recombine, emit light. The wavelength (or frequency, or energy) of the emitted light is a function of the semiconductor's energy band structure. If, however, the physical size of the semiconductor is reduced to quantum dimensions, additional energy is required to confine the electrons, which then shifts the light emissions to shorter wavelengths, which has an additional effect on . . . .

Disappointingly, there are no further color changes in the solution. Or rather, as the particles grow, atom by atom, their color's steady advance through the visible spectrum is occurring too slowly for me to see. "We grow them for hours," Vitasovic explains. "Depending on the color we want, we grow them up to 68 nanometers. That takes twenty-four hours." There's no shortage of colors to be seen around the lab, though; there are faintly glowing jars and vials and beakers everywhere. One small rack even shows off test tubes in each of the spectral colors: red, orange, yellow, green, blue, and violet. (Indigo is a spectral color as well, between blue and violet, but isn't represented here, probably because people generally can't distinguish it from blue. In this sense, most humans can be considered color blind.)

For essentially his entire career, Bawendi has specialized in these particles, which he calls "colloidal dots." My *Britannica* defines a "colloid" as "any substance consisting of particles substantially larger than atoms or ordinary molecules, but much too small to be visible to the unaided eye." Colloidal particles can be drops of liquid, but in the context of this research they are always solid bits of metal or semiconductor. Bawendi has explored their structure, tinkered with their composition, their dopants, their insulative organic coatings, and kicked around the physics and math that explain them.

Perhaps the most interesting thing he's done is also the simplest: drying them out in such a way that they self-organize. He calls the resulting crystals "artificial solids," and spends most of his time these days exploring their electrical and optical properties, which are strange indeed. One obvious question arises: why are these artificial crystals any different from natural crystals of cadmium selenide? The answer lies in the shape of the quantum dots and in the insulation of their organic coatings. Like the glass in fine crystal, this insulation prevents electrons from easily escaping the semiconductor. Also, the dots are usually spherical, meaning the contact area between them is very small—often just a few atoms. This also makes it harder for electrons to flow from one dot to the next.

These artificial solids can be amorphous (glassy) or crystalline. The nanoparticles can be not only semiconductors but also metals, and their organic coatings can vary in composition and thickness or even be stripped off altogether. Also, the particles are of remarkably uniform size, but they may be sorted using a precipitation process in order to obtain even more precisely uniform solutions. The shape of the particles is also remarkably consistent, and can be controlled to some extent by adjusting the exact type and mix of the organic chemicals.

In each of these cases, as Bawendi puts it, "quantum-mechanical coupling between adjacent dots is weak, and excitations are mainly con-

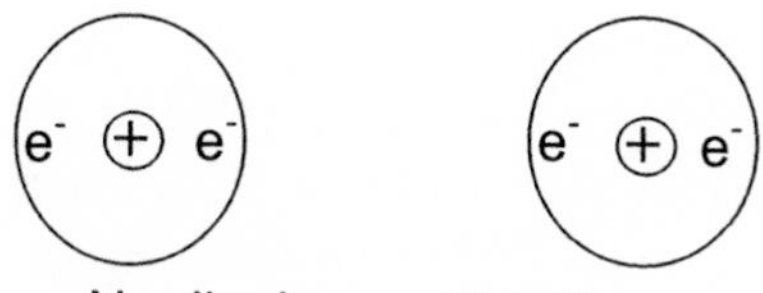

No dipole, no attraction

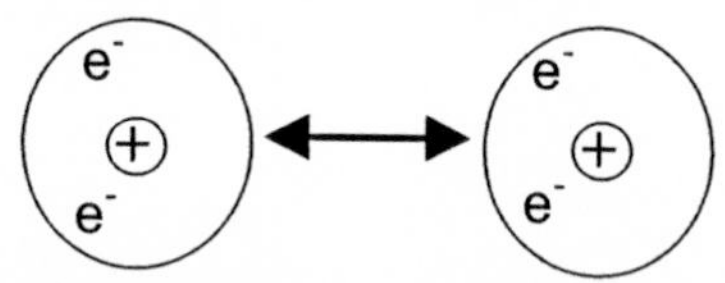

Slight dipole attracts neighboring electrons

FIGURE 3.1 THE VAN DER WAALS FORCE

The oscillation of electrons creates an intermittent asymmetry that draws quantum dots together.

fined to individual quantum dots." In layman's terms, the artificial atoms in the dots do not exchange electrons easily, and so are not attracted to one another by the equivalent of the covalent chemical bonds that hold many natural atoms together.

An attraction known as the "van der Waals force," however, can arise when neutral molecules, atoms, or particles develop an oscillating dipole, or small asymmetry in their electrical charge, due to the natural motion of the electrons inside them. (See Figure 3.1.) Opposite electrical charges are attracted to one another, while like charges repel, so even a very tiny asymmetry will affect the motion of electrons in neighboring particles, creating an imbalance in them as well, and causing all of the particles to be attracted together. This effect is weak compared to other chemical binding forces, but it's still important in the behavior of gases, organic liquids, and mineral solids composed of small, neutral particles. This is one reason clay particles stick together even when dry, while sand (with its much larger particle size) does not. The van der Waals force also affects Bawendi's colloidal solids—in fact, it's the only force holding the solids together at all. The analogy of oranges in a box winds up being less appropriate than, say, weak spherical magnets in a box.

So these solids hold together, but they're extremely fragile. They can be scratched easily, and they often spontaneously crack while drying. Interestingly, such damage doesn't appear to have a significant effect on their final properties. Bawendi explains, "These large crystals were an attempt to create an entirely new material, but we did not find any new properties related to its being an ordered crystal. Disordered solids behave the same."

So amorphous colloidal films, with a glassy, translucent appearance or even a scratched and cracked one, are just as good as colloidal crystals, and can be formed simply by pouring the quantum dot solution onto a surface and evaporating the solvent. The quantum dot powder can also be "drop cast" onto lithographically patterned substrates (such as silicon or sapphire) to further control their arrangement. Dozens of layers of this nanoparticle film can be placed on a surface, building it up into a genuinely three-dimensional solid. Alternatively, some types of quantum dots can be grown directly on a semiconductor substrate. On a lead telluride (PbTe) surface, molecules of lead selenide (PbSe) will spontaneously arrange themselves into 40-nanometer-wide pyramids. Layers of these two materials can then be stacked to produce a three-dimensional crystal. Other methods exist as well, and there's little doubt that many more will be discovered in the future. Thanks to people like Bawendi, artificial solids are officially a fact of life.

Unfortunately, the chemically produced dots in these solids don't have tiny electrodes running up to them, so they can't be controlled individually as artificial atoms. Still, by running a voltage through the entire solid, or through the substrate on which it was deposited, it's possible to drive electrons into and out of all the dots simultaneously. The substrate may even have metal electrodes laid down on it before the dots are deposited, so that the electrons have to travel through the nanoparticle film to get from the negative terminal to the positive. (See Figure 3.2.) Theoretically, it should be possible to lay down electrodes whose spac-

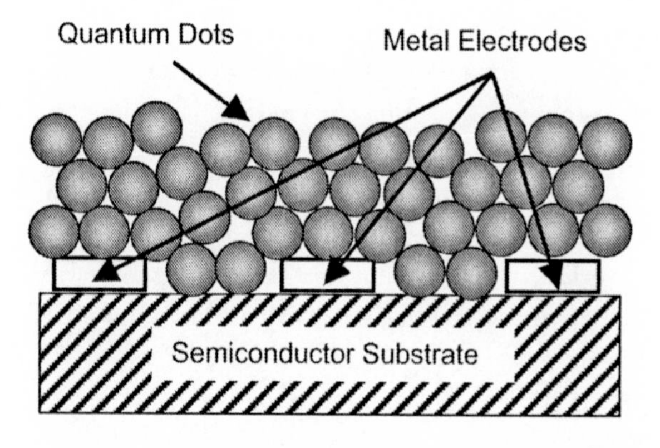

FIGURE 3.2 ARTIFICIAL SOLID

Bawendi's two-dimensional artificial solid is a rudimentary form of programmable matter. When electrons travel from a source electrode to a ground or drain electrode, they must pass through the colloidal quantum dot particles. This process excites the dots, increasing the number of conduction electrons and therefore the atomic number of the artificial atoms trapped inside.

ing is comparable to the size of the nanoparticles themselves, and to arrange for each particle to fall exactly between two electrodes, although technology doesn't yet allow us to try it.

Thanks to slight irregularities in size, shape, and number of dopant atoms for each dot, there's no guarantee that all these artificial atoms are behaving in precisely the same way. The average number of excess electrons can be controlled, though, and the resulting properties measured. Most of these properties, such as the rate of electron "tunneling" or teleportation from one artificial atom to its neighbors, are determined not only by the size and composition of the individual quantum dots but also by their spacing and by the properties of the intervening material.

Marc Kastner has worked very closely with Bawendi in this research, and their contrasting attitudes about it are, well, illuminating. Bawendi is interested in the particles themselves—their exact sizes and characteristics, while Kastner seems concerned mainly about the solid as a whole, and the wires connecting it to the outside world. Bawendi gets excited about spectroscopy—measuring the materials' characteristic emissions and absorptions of light—while Kastner wants to know about electrical resistance and capacitance. The two men complement each other nicely: they advance the field by working alone, but strengthen and consolidate it by working together. And they seem to agree that they're onto something, that artificial solids are slowly unlocking the dark quantum secrets of matter.

Bawendi has grown solids of very large size: tens of centimeters across and several microns thick. At the lower energy states, though, with few electrons on board, the dots are highly insulative, which unfortunately makes it nearly impossible to drive additional current through the solid unless the electrodes providing the voltage are very close together. Typically, the electrodes are tiny metal bars spaced 1–20 μm apart on the substrate, so the samples being tested are only a few microns across, often with only a single layer of dots atop the semiconductor. But with numerous electrodes laid down in patterns on the substrate, researchers can form areas of artificial solid with approximately the same excitation that are large enough to see with the naked eye, and to measure and probe with fairly ordinary instruments.

These experiments give us our first look at what programmable substances are like and what they may be capable of. They are the visible tip of a much larger iceberg, whose submerged mysteries we can only guess at.

One of the less surprising properties of Bawendi's artificial matter is fluorescence: very precise fluorescence at narrow frequency ranges. This also occurs in nature, but here the output frequencies are tunable with the size of the dot rather than fixed by the energy levels of an atom. Just turn the voltage source on, and away it glows. Bawendi's solids can also be excited optically rather than electrically: like the quantum dot solutions, they have the fascinating ability to drink in light at virtually any higher wavelength, and to spit it back out in a nearly monochromatic stream. The light source can be almost anything: white, colored, laser, ultraviolet. . . . What comes out is a single bright color determined by the exact characteristics of the quantum dots. These crystals also reflect, refract, and absorb light in interesting—and electrically variable—ways. Bawendi and his colleagues have investigated a number of fascinating properties with commercial and industrial potential.

Photoelectric Effects

The photoelectric effect, first observed in the 1880s and partially explained by Albert Einstein in 1905, occurs when photons strike a material such as a semiconductor or metal. The energy of the photons is absorbed by the electron shells of atoms, and as a result some electrons may shift from the dense-packed valence band (the insulative level where they normally reside) to the higher, looser energies of the conduction band. One consequence is photoconductivity, in which the conducting ability of the material increases with exposure to light—including infrared, ultraviolet, and x-ray light. This principle underlies many optical detectors, including the vidicon tubes of older television cameras.

There are also photoresistive materials, such as cadmium sulfide, whose conductivity *decreases* during illumination. Such materials are rare, though, and have properties such as bright color, luminescence (fluorescence with extended afterglow), and slow electrical response times, any of which may be undesirable. Photoresistivity has not been documented in quantum dot solids, but photoconductivity certainly has, and if the two effects can be produced together, or played off against one another, then a whole new generation of electro-optical devices may emerge.

A more important result, at least in technological terms, is the photovoltaic effect, which is used in solar cells for the direct conversion of light energy into electricity. This effect generates electron-hole pairs (i.e., knocks electrons off their parent atoms) in a material such as silicon, and if the electrons are forced to go in one direction and the holes in the other (e.g., at the asymmetric barrier of a P-N junction), then an electrical voltage is generated.

Today's commercial solar cells are around 13 percent efficient at converting sunlight into electricity, which makes them uneconomical in any but the sunniest climates. Even NASA's most sophisticated—and expensive—solar cells are usually no more than 24 percent efficient, although experimental multilayered designs have achieved upwards of 30 percent in the laboratory. (Notably, such converters once again blur the lines between a designer material and a collection of nanoscale devices.)

These efficiency numbers reflect a practical limit, not a theoretical one. The available materials—primarily silicon and other semiconductors—are simply not very photoelectric, and the junctions we can place in them are not very efficient electron-hole separators. With natural atoms, the choices are quite limited. But with designer materials, and particularly programmable materials that can adjust to changing conditions, there is every reason to believe higher efficiencies—maybe much higher—are possible. Geoffrey A. Landis, a specialist in photovoltaics at NASA's John Glenn Research Center in Cleveland, calls predictions of 50 percent efficiency "reasonable," although he dismisses any values higher than that as "speculative."

Reflection, Refraction, and Beauty

Many attempts have been made, with varying success, to link the concept of "beauty" to mathematical principles. One such correlation is the index of refraction for transparent materials; very high values tend not only to bend and distort the light passing through the material, but also to break it apart into pleasing rainbows. Absorption spectra, Bawendi would say: you get the full rainbow, minus the characteristic narrow frequency bands absorbed by the atoms in the material. Vacuum, by definition, has a refractive index of 1.0, meaning it does not bend light at all.

Air, with an index of 1.0003, is not much better. I have never heard anyone say that air or vacuum is beautiful—not to look at, anyway. But glass has an index of 1.5. You can make prisms out of it, or little horses or dragons or whatever. Sculpted glass is widely considered a beautiful commodity. Moving on up, we find leaded crystal and similar materials, which have been doped to increase their refractive index. These are more beautiful than glass, while gemstones, with still higher values, are considered more beautiful still. Diamond, with the highest refractive index of any natural material—a whopping 2.4—is the most beautiful of all.

Interestingly, in 2002 a team of researchers at the Air Force Research Laboratory in Massachusetts nailed down the far end of the scale by shining laser beams through a praesodymium-doped crystal of yttrium silicate, producing a highly excited material capable of slowing its internal speed of light to zero—equivalent to a refractive index of infinity. Light that enters the crystal simply stops, until the excitation source is turned off. Since there's nothing but blackness to look at when no light exits the crystal, one presumes the peak of beauty occurs somewhere between diamond's 2.4 and praesodymium yttrium silicate's infinity.

With their proven ability to manipulate a material's index of refraction, quantum dots will almost certainly find their way into ornamental objects. Harvard researchers have even made clear plastics into pseudogemstones by doping them with nanoparticles of metal. If such doping characteristics can be varied from one point to another in a material, it may even be fairly easy to create flat lenses that mimic the effect of convex, concave, or fresnel lenses through variations in the refractive index rather than in the thickness of the lens.

Another way to achieve beauty is to hide things people don't want to see. Windows are one example: if we can see them, it means they're dirty, which in turn implies that they're old or ugly or poorly maintained, or that the view outside is not worth seeing. Most artificial solids have a lower refractive index than their parent semiconductor, and if these values can be nudged lower, it might be possible to create "invisible" objects with optical characteristics similar to air or vacuum.

Still another form of beauty is found in pure metals, which tend to be lustrous. This is another way of saying that they reflect light very efficiently (i.e., that a photon striking a metal atom rebounds at a complementary angle, with little change in its energy). Some metals, such as gold, absorb certain frequencies while reflecting others; this gives them highly distinctive colors in addition to their luster. A few elements are good reflectors for all wavelengths in the visible spectrum. This is especially true when they're molten—think of a drop of quicksilver—but a few of them work well when applied in thin layers to another material. This is how mirrors are made.

Mercury has a reflectivity ranging from 79 percent in the ultraviolet to 90 percent in the near-infrared. Silver is much better; its reflectivity is rather poor in the UV, but it reaches 98 percent in the visible and 99.5 percent in the infrared. Aluminum is nearly as good. Other notably shiny metals include sodium, potassium, and chromium. No mirror is perfect, though; at every frequency, at least a little bit of energy is absorbed. When large amounts of light are being reflected, as in a high-powered laser or solar oven, the buildup of heat can present major challenges. It is conceivable that quantum dot materials will provide us with better reflectors than any natural atom.

More interestingly, the ability of artificial atoms to hold an asymmetrical shape may make one-way mirrors a genuine possibility. Today's "privacy glass" is really just a half-silvered mirror, designed to reflect 50 percent of the light that strikes it and to transmit the other 50 percent through (minus efficiency losses, of course). The "privacy" effect occurs only when there is a strong difference in illumination levels from one side of the glass to the other. A person standing in a bright room sees 50 percent of that light reflected back at him; someone in a dark room sees 50 percent of his own dark reflection, but it's overwhelmed by the 50 percent coming in from the bright room. This is why skyscrapers are mirror-bright by day but seemingly transparent at night, when their internal lighting is brighter than the night sky. You would not see this effect in a material that behaved like a metal on one face and a transparent insulator on the other.

Quantum Color

Another property exhibited by some of Bawendi's artificial solids is color change—specifically, darkening—when certain conditions arise. "Electrodarkening" occurs when a voltage is applied across the material. This effect is also seen in liquid crystals and should be familiar to anyone who's ever owned a digital watch or observed a laptop's LCD screen. By itself, the effect can achieve only shades of gray and black, but if the electrodarkening material is placed in front of a color mask and a backlight or back-reflector, then dark areas will block the transmission of a colored pixel, while light (transparent) areas permit it. This allows a laptop screen not only to change color but also to display moving pictures.

The major advantage quantum dots provide is that they're three to four orders of magnitude smaller than the thin-film transistors of a laptop screen, and may in fact be smaller than a wavelength of visible light. And since they can glow as well as darken, they may also be able to replace the functions of the backlight and color mask. K. Eric Drexler's Foresight Institute, a nanotechnology think tank, has used the term "video paint" to describe these and related possibilities. This description may be a bit extreme and oversimplistic—you'd still have to pass control signals somehow to the display's trillions of individual pixels. This is a real challenge, and any system or device that attempted it wouldn't be "paint" any more than a TV screen is paint. Still, Drexler's language is evocative.

Photodarkening, another form of color change, is also familiar: it's what light-sensitive sunglasses do. They do it slowly, though, and you have limited choice of material (usually a metal-doped glass) and of color (the most common being an unpleasant blue-gray that does little to enhance fine details). And the material is never fully opaque or fully transparent. In contrast, the darkening response of artificial solids can be effectively instantaneous, and their other characteristics are far more flexible than anything present technology can offer. These solids may someday be cheap enough to put into windowpanes, perhaps pressed between two sheets of ordinary glass, to help regulate building temperatures on sunny days.

The opposite of photodarkening—what Bawendi calls photoreflectivity—occurs when a material becomes more reflective in the presence of increasing illumination. This, too, is helpful in temperature regulation, especially in environments like deserts, where bright light is inconvenient or even hazardous. Photoreflectivity might be triggered by visible wavelengths, but would be even more useful when triggered by thermal infrared light, or "radiant heat." The ideal would be a material that was an infrared transmitter or absorber at low temperature and an infrared reflector or one-way mirror at high temperature.

A closely related area is thermochromicity, in which a material's color is indexed directly to its temperature. A familiar thermochromic material is liquid crystal, which is used in flat skin-contact thermometers, bath-water thermometers, and "mood rings." Thermochromic plastics are also popular in baby baths, where even minor temperature variations can be a problem. White plastics that turn red above an upper threshold temperature, and blue below a lower threshold, are increasingly available. Chinese researchers have even developed a thermochromic paint that reacts in the opposite way, turning a cool blue shade when warm, a warm red shade when cool, and a pale green at the perfect room temperature.

This is done mainly for thermal regulation—the paint can increase the temperature of a building by about 4°C in winter and decrease it by about 8°C in summer—although the researchers also claim an aesthetic benefit to having a home's color match the season. Even more impressive effects could be achieved if black and white (or black and silver) were available color choices. Alas, these are not only less aesthetic but also unachievable with present-day thermochromic materials. But it isn't hard to imagine a quantum dot material with these characteristics, and even if this were difficult to produce, the same effect could probably be achieved with "video paint" and a digital thermometer.

The glass we use today is a silica-soda-lime (SiO_2, NaO, CaO) preparation that is wonderfully transparent to visible light but unfortunately quite reflective to long-wavelength infrared, a.k.a. radiant heat. This reflectivity creates the well-known "greenhouse effect," which is not always desirable. If a similar material could be made switchably transparent at these wavelengths, it too would go a long way toward regulating the temperatures of buildings.

In related work, researchers at London's Imperial College of Science have created an artificial solid that is normally opaque but turns transparent when excited by a laser. Like many other quantum-dot devices, this one is intended for use in quantum computing, which may be far easier to accomplish with photons than with electrons. I personally have a hard time getting excited about quantum computers, though, because they seem so much like ordinary computers. Enormously more powerful, yes, but so what? Calculating machines of one sort or another have been around for centuries, and Moore's Law has already desensitized us to rapid changes in computing power. Better computers are always welcome, but for me the really exciting ideas are the ones that involve sudden and dramatic changes in the properties of matter. These get us away from Moore's Law entirely, and into the realm of Clarke's Law. Such technology has few if any analogs in the classical world, or even the twentieth-century one.

One final electro-optical trick is suggested by Howard Davidson, a Distinguished Engineer and "quantum mechanic" on Sun Microsystems' corporate staff. He points out that the frequency of a laser (i.e., its color) depends on the gain frequency of the material generating it—an electrical property—and the resonant frequency of the mirrored chamber in which the beam accumulates. Typically, the resonant frequency is adjusted by physically stretching or shrinking the chamber—something you could potentially do nanomechanically or with tiny piezoelectric effects—but with quantum dot materials you could also do it by adjusting the index of refraction. Balancing electrical and optical characteristics this way, while preserving the lasing properties of the material, would be a delicate art. But the result would be one of the holy grails the photonics industry has dreamed of for decades: a variable-color laser.

Just for starters, this innovation would permit optical networks to carry vastly more traffic, and could also lead to adaptive communication lasers whose wavelength varies with atmospheric conditions. In conjunction with photovoltaics, it could also be useful in the direct, point-to-point transmission of energy.

The optical properties of quantum dots are not only a function of their size and composition. They also exhibit variations with temperature, applied electric field, excitation energy, intensity of illumination, separation distance between particles (e.g., by an insulating layer), and the exact surface properties of the insulating material. Some theories even describe ball lightning as a quantum dot phenomenon, as lightning strikes vaporize the soil and form interacting nanodroplets of pure, highly excited silicon in the air. At this point, science has really only scratched the surface of what may eventually be possible.

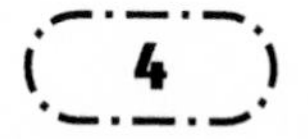

Thermodynamics and the Limits of the Possible

> The quantization of charge on a natural atom is something we take for granted. However, if atoms were larger, the energy needed to add or remove electrons would be smaller, and the number of electrons on them would fluctuate except at very low temperature.
>
> —Marc Kastner (1993)

HARVARD UNIVERSITY, also in Cambridge, Massachusetts, also nestled against the north bank of the Charles River, is as beautiful and inviting as MIT is utilitarian. The two campuses are separated by a mile of residential housing—largely faceless—but as Harvard approaches, the buildings are suddenly all in the colonial style, with red brick and white mortar, white windowframes and white wood trim everywhere. There are chimneys topped in copper, and towers with colorful domes topped by weather vanes of brass. The sidewalks are broad and lined with stately trees.

MIT is a concrete kingdom with clearly defined borders, but here the campus blends into the surrounding neighborhood, with little restaurants and shops and cafes and other businesses. Walls and gates ostensibly mark the university's boundaries, but the buildings outside the walls look just like the ones inside, and even the ones that aren't Harvard appendages are bits and pieces of Radcliffe or the JFK School of Government or the dozens of religious schools and prep schools that dot the map like shotgun scatter. This is the schooliest neighborhood of the schooliest suburb of Boston, which would itself be one of the world's great college towns even with Cambridge removed.

Harvard's architecture is homey rather than monumental—one has to fight the impression that even the largest buildings are overgrown houses, and not the libraries and lecture halls of a world-renowned brain shop. Even in midsummer, the area around the campus is crowded and vibrant. And the crowd is hardly a lazy one—there are paved trails along the Charles River, and the streets are full of bicycles and joggers. The overt signs of wealth here, while detectable, have the quiet ease of very old money. This is simply the place to be, and one imagines people huddled all around, living as close to this neighborhood and its styling as they can possibly afford to. Others, pragmatically numb to its charms, are content to squat in whatever miserable hovel suits their immediate purposes, as any long stroll through Cambridge will tell you.

The Lyman Physics Building could easily be mistaken for an apartment complex, condominium, or brownstone office space. It's very small, which may give some indication of the relative importance of physics here in the land of economy and politics, law and liberal arts. It was also the first dedicated physics lab ever constructed in the Americas, and was built entirely without iron nails. This was supposed to facilitate research into electricity and magnetism, but then it turned out that the building's brick facades drew their red color from a hefty dose of iron oxide, a ferromagnetic mineral more commonly known as rust. Oops. It's a kind of fable of American science. Still, Lyman has been the site for numerous breakthroughs in the fields of optics and acoustics.

It is also the construction site for Harvard's new Center for Imaging in Mesoscale Structures (CIMS), whose leaders include a Dr. Charles Marcus, recently imported for this purpose from the hallowed, earthquake-rattled halls of Stanford University. It says a lot, I think, that all three of his graduate students made the journey as well.

Marcus—Charlie to his students—is something of a Johnny-come-lately to the field of quantum dots. In modern science, a researcher is expected to choose either the path of the theoretician or that of the experimentalist. Marcus remained ambivalent, postponing the choice well beyond any seemly interval. Only after his postdoc assignment did he discover the wonders of nanostructure, and decide finally that he was "not smart enough at theory" to be a theoretician of the highest order. He did, however, have good experimental skills, including mechanical aptitude, a finely honed aesthetic sense, and a knack for "decision tree winnowing" to rapidly determine "all the things a phenomenon isn't." This last—as anyone in the fields of science and technology could tell you—is a particularly rare skill, and it explains his rapid rise in the world of quantum confinement.

"I came to nanostructures as an outsider," he says with a grin. "I was interested in neural nets and chaos theory, and in the relationship between quantum theory and chaos. I had no prejudice about what was impossible." In an age of specialists, Marcus has a touch of the Renaissance man about him. He is, among other things, working personally with the architects of the new mesoscale physics wing, to ensure that form and function mesh with proper elegance. "I'm reading books, sitting on the committee. . . . I'm fascinated by the tradeoffs. This is one of the things I like about Harvard: the opportunity for breadth."

The same *elan* shows clearly in the design and decor of Marcus' office, which looks like it should belong to an upscale architect or CEO. First of all, for a professor's office it's quite big. The computer equipment is brand-new: there's a flat, very large LCD screen on the desk, with a Macintosh logo emblazoned across its top. The CIMS, it turns out, is a hotbed of the MAC/PC wars. The thing you really notice, though, is all the beautifully varnished wood, the deep, cream-colored carpets, the white walls, the halogen track lighting. The decor is the exact opposite of institutional.

"The recruitment honeymoon includes a remodeling budget," Marcus says. "The first thing I did was throw away the catalog they gave me and pick up a phone book. I found this low-cost Mom and Pop woodworking shop that seemed to understand what I wanted, so we worked together on a custom design. I never plan to leave, so I figure it should be something I really like."

Marcus, perhaps the youngest of the field's senior researchers, is always busily in motion, always on an errand, always engaged in jovial conversation with someone, with everyone. Laughing about it, he tells me, "Why bother hogging glory when there are so many hard problems still to be solved? The greatest threat out here is dying of loneliness." While we talk he is literally holding meetings with other people, monitoring various experiments, and showing off a bottomless thirst for coffee. One of the errands we take is to the local beanery, where they know him by name, and know his preferred blend without asking.

He seems barely older than his graduate students; they josh together casually, everyone seems to be having a good time, and yet there's really no question at all who's in charge. Marcus exudes the same quiet excitement as the other scientists. He clearly feels that his work is both important and rivetingly exciting. "This job," he says, "is not a grind."

A conversation about the work going on here is, fundamentally, a conversation about liquid helium. Why? Because the energy of the electrons in a quantum dot is a direct function of the dot's physical size. Ten nanometers (the size of Bawendi-type dots) is a magic number only because it equates to the de Broglie wavelength of a room-temperature electron—meaning it can produce meaningful quantum effects without refrigeration. Larger dots can be constructed to house electrons with lower energy, and smaller ones can contain higher energies. And higher energies are good, because the energy levels in an atom are supposed to be quantized into discrete, well-separated bands. This occurs more and more easily, the higher the energy is.

But lower-energy bands are blurrier and closer together—less like atoms, more like man-made electronic devices. This is bad, because heat is energy, and "thermal noise," the random transfer of energy and vibration between atoms, can excite the atoms' electrons just as easily as light rays or electrical voltages can. But the size and timing of this excitation is random—a statistical phenomenon. So if the gap between two energy levels in a quantum dot winds up being smaller than the error bars of thermal noise, the electrons will freely hop from one energy level to the next, and back again, at random intervals. For practical purposes, this means the energy levels are not quantized at all but continuous. The electrons trapped in the device will behave classically rather than quantumly, and they will be unable to form an artificial atom. Instead, they'll jumble around in random clouds, just as they would in a bulk conductor or semiconductor—a phenomenon Marcus refers to as "thermal smearing."

If the energy levels in the device cannot be raised (e.g., by making the quantum dot smaller), then the only cure is to reduce the thermal noise. When this noise is much smaller than the natural spacing of the device's energy levels, then quantum mechanics takes over again, and atom-like behavior is restored. For this reason, today's large quantum dots—some of them a thousand nanometers or more in diameter—display interesting behavior only at cryogenic temperatures.

This supercooling is accomplished in a device called a "dilution refrigerator," which uses the evaporation of liquid helium to maintain temperatures below 1 Kelvin—colder than interstellar space, where the afterglow of the Big Bang keeps things at a balmy 3 Kelvins. The earliest of these refrigerators ("boringly called helium cryostats," says Marcus) used ordinary helium, a.k.a. helium-4 or ^{4}He, which can be extracted in small amounts directly from Earth's atmosphere. It's produced by the radioactive decay of uranium deep inside the planet, and seeps upward through the crust. Helium-4 is abundant in subterranean deposits of natural gas, particularly in the United States, where it constitutes as much as 8 percent of the total gas. So it's not expensive stuff, even in bulk, even in its liquid form. About $50 a kilogram.

For high-end quantum dot research, though, the 1.2 Kelvins of evaporated ^{4}He are still way too hot. The artificial atoms in larger quantum dots quickly become artificial ions, and finally artificial plasmas.

Fortunately, another isotope of helium, called helium-3 or ^{3}He, is lighter than ^{4}He and has a lower vapor pressure, which makes it a superior refrigerant. Unfortunately, it's extremely rare in nature, and is produced commercially only by nuclear reactors. Filling a fridge with this stuff costs nearly ten thousand times as much—potentially hundreds of thousands of dollars per experiment. Fortunately, the optimal mix—capable of producing temperatures as low as 0.01 Kelvin—turns out to be a small amount of ^{3}He dissolved in a bath of ^{4}He.

This fluid, a witch's brew if ever there was one, is piped, hissing, into the dilution fridge, which boils with fog. When the hissing and boiling finally subside, there's a liquid reservoir of the stuff at the bottom of the dewar. Electrostatic quantum dots, etched into chips like the ones Marc Kastner uses back at MIT, are wired up, mounted on a high-tech dipstick, and lowered into this reservoir, which boils anew until the lid is sealed. The entire apparatus is suspended with bungee cords—the best vibration isolator anyone has yet devised for it—and finally enclosed in a Faraday cage, a metal-lined room that prevents the intrusion of electromagnetic fields. Even stray radio signals, absorbed by the antenna-like wires feeding down to the quantum dot, could heat things up enough to destroy the subtle effects being observed.

In one experiment, Marcus puts his hand on a knob and turns it, watching a gauge of red numbers. "I'm adjusting the conductivity of a material," he tells me, shaking his head as if in disbelief. "It's still amazing. These measurements are purely statistical, by the way. It takes an understanding of quantum chaos to extract the useful parameters."

In the background, working and talking, is a group of young men and women. The field's literature makes it seem predominantly male, which probably reflects the makeup of chemistry and physics programs fifteen and twenty years ago. But about half the graduate students today are female, so the field's future is certainly a mixed gender enterprise. There have even, I understand, been a few marriages. At the moment, though, the discussion is work-related, concerning a student in another lab who has been showing great aptitude as an experimentalist. "And he's an undergrad!" someone exclaims. "He's just here for summer internship!" Marcus looks up, nodding absently. "Yeah, you never know. Some people just never pick this stuff up, no matter how much you train them. Others take to it right away. Let's get this guy on board."

Steal him, in other words; hijacking student workers is a time-honored tradition in academia. Still, it rarely happens without the student's own consent, so I can't help asking: "What makes you think you can get him?"

Marcus laughs off the question. "Oh, we can get him. There are limited areas in physics where true, basic research also has clear—and potentially huge—implications for the real world. Students are drawn to that. I may lose students to Kastner, but I won't lose them to string theory." Then he's off again, talking to someone else about something else, about using a layer of liquid nitrogen to extend the lifetime of fridge experiments to six days. "Six *days?*" he asks, clearly impressed. "Yeah, definitely, let's find out about that." And then he's off again on something else.

From a technological standpoint, all this helium stuff is of course a dead end; any products that emerge from it—such as the quantum computers our military is hoping for—will be bulky, expensive, and dependent on massive infrastructure. These dilution refrigerators are not money makers but money sinks. Through the eyes of a physicist, though, the world is not so much a race to market as a cooperative logroll to wisdom. The important thing is to understand phenomena like quantum confinement and decoherence, and understand them *soon.* The quantum dot is a piece of lab equipment, no different from a microscope or a Bunsen burner.

Sometimes understanding is advanced through solo missions or the thought experiments of a theorist; more often it happens through big collaborative projects built on huge bibliographies of prior research and established technique. For the moment, making helium baths colder is an easier trick than making nanostructures smaller, finer, and more precise, so that is the direction a smart physicist will go. If other, more promising avenues open up in the future, then Charlie Marcus and his colleagues will flow that way, sure as water flows out through a leaky dike.

The difference between engineering and science has never been clearer to me than during this conversation. We engineers want *things*. Engineers see things as an end goal, and science as a means of achieving that goal. Engineers are lazy, too; we want equations and explanations handed to us in textbook form, for immediate use, with or without real understanding. Physicists are more curious, more thorough, more patient with details. But they have their own sort of laziness: never worrying about the best path, the most useful path, the path with the highest ultimate payoff. They're not unaware of it—the idea just doesn't excite them.

"I'm not a product developer," Marcus acknowledges. "I'm a scientist. You can have all my money. I need a house for my family, and some food, and a place to do my work. That's about it."

Well, maybe. He grew up in the wine country of California, and he's married to an artist who grew up there as well. They have two small children together, and they live right here in Cambridge, a few blocks from the university. Aesthetics matter to Charlie Marcus. The future matters. His favorite pizza place matters. He's oversimplifying to make a point, and the point is valid, but there is more to it than that. Harvard isn't known for its physics labs—not nearly to the extent that MIT is—but everything here is newer, nicer, more plentiful. And that's in the *old* building.

When badgered, he relents a bit, but also digresses. "These very cold temperatures are a problem, but also an opportunity. The important quantities in physics are all dimensionless. Energy scales with size, and we've got devices all the way up to 8 microns. We want to remove size from the equation. I'd love to say that temperature doesn't matter at all, that these phenomena scale cleanly. But for one thing, the movement of electrons in quantum dots shows relativistic effects. Also, impurities in the structure create electric fields, which look like magnetic fields to a moving electron. So the magnetic moment of an electron precesses when these impurities are present. That's very hard to control at small sizes, although it may have applications."

He thinks about it for a moment, then continues: "What we really want today are quantum dots with as little personality as possible. Symmetry produces shell structures, like an atom, but atoms are a special case. They have personality rather than generality. To learn the fundamentals, we want stuff that's asymmetrical, so their behavior is driven purely by the electron energies. This is easier when they're large. Not circles, though; we want shapes which produce nonintegrable classical motions. Hockey rinks. If the classical behavior is chaotic, then we find that the quantum behavior is universal."

Marcus is referring to a well-known principle in chaos theory: the motion of a hockey puck in a rectangular or circular arena is easily integrated—the puck's position at any future time point can be predicted with simple, closed-form equations. But if the circle is sliced in half and placed at the ends of a rectangle—producing the elongated pill shape of a standard hockey rink—then something quite different happens: the puck's motion becomes chaotic. Not random—a detailed simulation of the puck can still predict its motion accurately. But a closed-form equation cannot. In chaotic systems, every event is dependent, in a wild and sensitive way, on the events immediately preceding it. A pencil rolling on a desk is classical, but one balanced on its point is chaotic: certain to fall, but uncertain as to how or when or in which direction.

The "hockey rink" description turns out to be literal; Marcus has actually worked with quantum dots that have this exact shape. (See Figure 4.1.) He puts one under a microscope for me to look at. At the lowest magnification, the scenery is all familiar: the insectoid chip carrier, and inside that the chip itself, webbed in by tiny gold threads. The chip looks much as you'd expect, much like any other chip you've seen up close or in pictures, except that there are gold traces raining down into its center like the spokes of some strangely shaped bicycle wheel. These are the photolithography traces. At still higher magnification I can

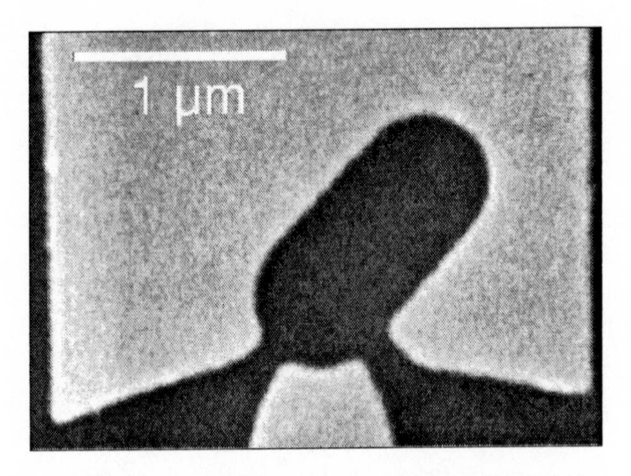

FIGURE 4.1 "HOCKEY RINK"—SHAPED QUANTUM DOT

Thanks to the physics of chaos, this shape turns the classical trajectory of a confined electron into a complex tangle that resembles, in some ways, the quantum waveform of an orbital. (Image courtesy of Charles Marcus.)

see, sprouting from their ends, even finer traces defined with an electron beam. And where these come together, at the very center of the chip, is the just-barely-perceptible shape of the quantum dot itself: a hockey arena with boards of gold.

"We might still dispense with temperature," Marcus tells me, "except that decoherence is strongly correlated with it. At 1 Kelvin, our decoherence time is extremely short—100 picoseconds. Colder, at 100 millikelvin, it jumps to 1 nanosecond, which is comparable to the clock cycle of a 1-gigahertz computer."

Long enough, in other words, to be caught in the act, before the wave functions collapse into classical particles. Long enough to be computationally useful. And at lower temperatures, decoherence can be staved off for even longer. "For electrons," Marcus notes, "the Pauli exclusion principle tells you that if low-energy states are filled up, they're off-limits to another electron. That's the origin of the effect. The temperature dependence of dephasing is from the fact that there's nowhere for the electrons to scatter to when all the states are full."

As another physicist, Ken Wharton, explains, the temperature correlation may also depend on "the sheer number of interactions between the quantum state and the rest of the universe. In general, higher temperatures tend to equal more interactions, because there are a lot more blackbody photons emitted from hot surfaces, which can then be absorbed and destroy atomic superpositions. But photon-photon interactions have such a low cross section you don't have to worry about it for optical quantum states. A photon that's in a quantum superposition is therefore going to be a lot more stable at room temperature."

This may be one more reason to downplay the application of chilled quantum dots to the field of computing; optical quantum computers appear to sidestep all the messy thermodynamics and simply function at room temperature. Marcus cautions, though, that these effects can't be produced by ordinary linear optics, but rely instead on the nonlinear responses of certain exotic materials—including quantum dots. So the primary advantage of quantum dot computers may not be in qbits or single-electron transistors, but in the near-instantaneous reconfiguring of optical and electrical properties. We'll return to this concept later; the point is simply that decoherence, while important, affects only a small subset of quantum dot capabilities.

As Bawendi's work clearly shows, the world has many potential uses for quantum confinement and atom-like behaviors at higher temperatures—a point Marcus happily concedes. "There's a lot of interest in buckyballs and nanotubes for this reason: they work in the quantum domain at room temperature. They have a low buy-in cost, and they're easy to fool around with. They're interesting because they're very small, yet very easy to make."

The term "buckyball" is short for "buckminsterfullerene," a form of carbon first discovered by Richard Smalley and Robert Curl at Rice University in 1985. The molecule, about 1.4 nm across, contains sixty carbon atoms arranged in a spherical pattern of pentagons and hexagons that is, by sheer coincidence, identical to a soccer ball. In mathematical terms, this shape is a truncated icosahedron, but its more popular name is an homage to architect Buckminster Fuller, who used it (and related shapes) for the construction of domes from flat polygons. But C_{60} turns out to be only the material's stablest form; the same chemical and physical processes can produce larger or smaller molecules in a variety of shapes, collectively known as "fullerenes." Nanotubes are a specific form of fullerene, essentially a buckyball that has been cut in half and used to cap a graphite cylinder of arbitrary length. (See Figure 4.2.) By the late 1990s, a powder of purified, single-walled nanotubes, vaguely resembling black felt, was available on the Internet for as little as $200 a gram.

There is considerable evidence that fullerenes can act as quantum wires and quantum dots. This is not surprising considering their small size, and considering that carbon can, under some circumstances, act as a semiconductor or metal. (In fact, as a conductor it supports much higher current densities than normal metals do, and sometimes even displays superconductivity at liquid nitrogen temperatures, or perhaps even higher.) A buckyball is different from a Bawendi-type CdSe dot of equivalent size, though, because it's hollow. Every atom in the ball is a surface atom, and every electron trapped on it will be trapped on its surface rather than bouncing around in its interior. This makes their actual behavior hard to predict or model with current techniques, while their small size makes it extremely difficult to attach wires to them for any sort of direct measurement. What we know about their properties generally comes from monkeying with bulk materials rather than individual molecules.

There are also "dopeyballs" and "dopeytubes," which have other atoms trapped inside the fullerene cage, and these materials have properties that are even more different and even more complex. But in the end, it may not matter; trapping electrons in a quantum-sized space, at quantized (i.e., discrete and well-separated) energy levels, may produce something sufficiently atom-like to fool even Mother Nature. The

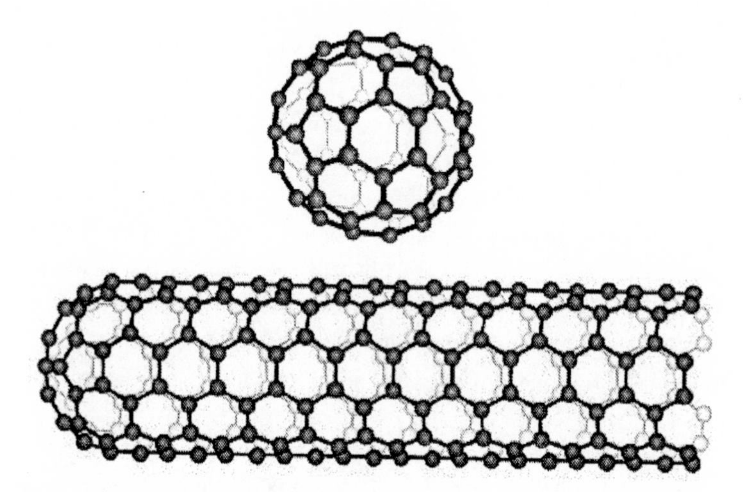

FIGURE 4.2 BUCKYBALL AND NANOTUBE

These intricate molecules, made of pure carbon, display interesting quantum properties even at room temperature. (Image courtesy of Richard Smalley and Rice University.)

chemical, optical, and electrical behavior of real atoms is defined mainly by the outermost layer of electrons anyway, so buckyball "atoms"—a tenth the size of Bawendi's dots and chemically all but indestructible—are a real possibility, even at room temperature.

This is not to say it will be easy. "There have been some breakthroughs," Marcus says, "in controlling the shape and properties of fullerenes. We know, more or less, how to sift out the ones that behave like a metal or an insulator or a semiconductor. But try making a simple Y-junction. Try making a loop. Nanotubes are found naturally in soot; we know exactly how to make them. But manipulating them is something else altogether. They're interesting, but right now we don't know how to use them."

In 1997 the research group of Paul McEuen (then at Berkeley, now at Cornell) produced crude single-electron transistors from carbon nanotubes. These functioned for only a few minutes, and only at cryogenic temperatures. Then, in 2000, Cees Deckers' group at the Delft University of Technology in the Netherlands produced a nanotube SET that functioned at room temperature, and in 2002 IBM built a nanotube transistor with a higher current density and a faster switching rate than even the best MOSFETs. Nanotube diodes have also been demonstrated. Still, these are all linear devices, with no branching and no looping-back. At the time of this writing, the only electrical circuit made from them relies on mechanical flexing of the tubes to make and break connections. Marcus' point is well taken: even if solid-state nanotube circuitry is theoretically possible, we may find it's too difficult or too expensive for us to produce.

Then again, we may not need to. Could the same effects be achieved in a semiconductor?

Back to Electrostatics

The energy of electrons and other particles is measured in "electron-volts" (eV), where 1 eV is equivalent to 1.6×10^{-19} joules, or 214 horsepower for a trillionth of a trillionth of a second. Marcus informs me that for pancake elements in a two-dimensional quantum dot, the spacing between the first two electron energy levels is an inverse function of the dot's area. For a 1-square-micron dot, the energy is 7 millionths of an eV, and the energy is ten times as much for a dot with one-third the diameter. These energies can be divided by a number called Boltzman's constant (86 meV/K) to yield an "ionization temperature" in Kelvins. Above this temperature, the electron will be energetic enough to escape from the dot. Using this rough approximation, we can compute that a circular artificial atom becomes stable at room temperature when its diameter is around 20 nm—no coincidence, since this is very close to the de Broglie wavelength of a room-temperature electron. This doesn't mean the artificial atom will behave completely and totally like a real atom, just that thermal vibrations won't be enough to kick away its electrons.

An atom twice this size would absorb and radiate only infrared light. To exhibit a "color" in the visible spectrum, the electrons need to have energies comparable to those of a visible photon. This occurs, for excited states only, when the atom is between approximately 10 nm and 30 nm—the size range for Bawendi's colloidal dots. But real atoms are 24–500 times smaller than this, and the outer-shell electrons that define their chemical, optical, thermal, magnetic, and mechanical properties have energies between 0.5 and 2.0 eV. To get truly atomic behavior, we need to create artificial atoms the size of natural atoms. This may be impossible.

Can we come close? According to the approximation given above, any dot smaller than about 2 nm should be capable of trapping a 2eV electron. Unfortunately, Marcus cautions that the uncertainty in his measurements makes this calculation imprecise. Still, we do know that very large natural atoms such as polonium and francium, about 0.5 nm across, behave in a satisfyingly atom-like way. They have a definite color, a definite chemistry. They're real atoms. We also know that 1.4 nm buckyballs and nanotubes exhibit definite quantum behavior at room temperature, including absorption and emission (i.e., color) in the visible and ultraviolet spectrum. So maybe 2 nanometers is a reasonable estimate after all.

Making a 2 nm electric fence on top of a quantum well sounds tricky at best, but fortunately, an artificial atom inside an electrostatic quantum dot is smaller than the fence enclosing it. In fact, the more charge you put on the fence's electrodes, the more tightly the atom will be compressed. (See Figure 4.3.) Anyway, the atom's size is somewhat hypothetical since its boundaries are statistical rather than physical. It's like trying to find the edge of a fog bank or the end of a flashlight beam. We do know that small is good and versatile, and that smaller is better and more versatile. In fact, by erecting *concentric* fences atop a quantum well, we should be able to create quantum dots with tunable sizes and energies, which are more versatile still.

This arrangement—an artificial atom of variable energy *and* shape—could be further enhanced by placing a second, identical set of

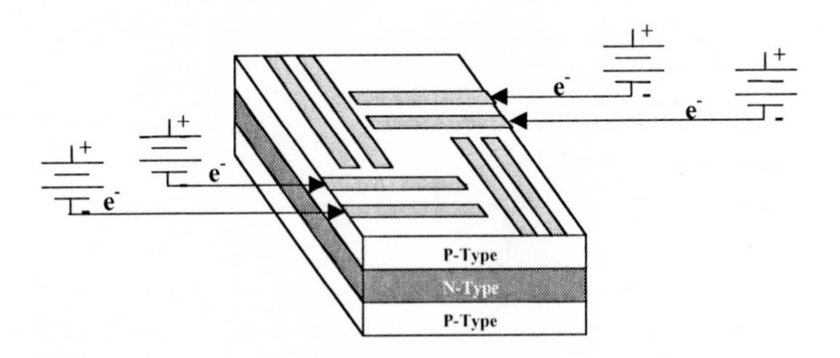

FIGURE 4.3 IMPROVED ARTIFICIAL ATOM

With eight voltage sources, this electrostatic quantum dot device allows precise, two-dimensional control over the artificial atom's size, shape, and atomic number. Adding eight more electrodes on the P-N-P junction's bottom face allows these adjustments to be made in three dimensions.

electrodes on the quantum well's lower face, probably with an insulating layer below that. The atom's characteristics would thus be fully adjustable in three dimensions. This is important because pancake elements (see Figure 2.7), lacking one whole dimension to cram electrons into, have a much simpler periodic table than spherical elements. If you want the full richness of nature—and more!—you really need that third dimension.

Even more complex arrangements may be possible, but these structures are so small—made of a few thousand atoms at most—that we don't have the sort of broad flexibility we do at larger scales. It's like building with Lego blocks—the smaller you get, the more restricted your design choices. Anyway, without adding any additional complexity, the device described above would be capable of introducing artificial "dopant" atoms near the surface of a semiconductor. In addition to the optical and electrical effects already discussed, such alteration of the semiconductor could dramatically affect its thermal and thermodynamic behavior as well.

Energy Storage

The three laws of thermodynamics, in their simplest form, are (1) you can't win, (2) you can't break even, and (3) you can't get out of the game. Energy is never free. Quantum dots can't break these laws any more than other electronic devices can. They're not able to change their mass, and since the devices themselves can't change their shape, their inertial properties are fixed as well. They can't function without a power supply and a place to dump their excess heat; pumping a 1 eV electron into a quantum dot takes a minimum of 1 electron-volt of energy, and letting it out again will release that energy, most likely in the form of waste heat. These are the facts of life that forever separate technology from magic.

But while energy can neither be created nor destroyed, it can be moved around, changed in form, and also *stored*—for example, in batteries. The most common batteries for reusable energy storage are lead-acid and nickel-cadmium (NiCd) batteries, which unfortunately run only about 65 percent efficient. In other words, 35 percent of the energy you put into them is converted to heat and irretrievably lost. These batteries are also massive: up to 30 kilograms (66 lb) to store a single kilowatt-hour (kWh) of energy. More recently, nickel-metal-hydride (NiMH) and lithium-ion batteries have achieved efficiencies of up to 90 percent and densities of under 10 kg/kWh. Although they were developed for cellphones and are not yet available for bulk storage, they do point the way to the future.

For contrast, a kilogram of gasoline holds about 12 kWh of releasable energy—a hundred times better than today's best batteries. Batteries are also limited in the number of times they can be charged—at some point, they short out and simply start behaving as resistors. This is one reason we continue to burn gasoline in our cars and trucks rather than convert them all to electricity.

Still, there is another means of storing electrical energy that has long been used in electronics and is finding its way into automotive and other applications. A battery stores energy chemically, in the form of reactive ions, but static electricity is an older and perhaps more familiar way to store it. Shuffle your feet on a carpet and you will accumulate a substantial electrical charge. Touching a pipe or doorknob then forms a connection to the ground, through which your charge can escape, often with the painful crack of a high-voltage electrical arc.

There is an electronic component—the capacitor—that does exactly the same thing in a more controlled way. A capacitor is simply a pair of conductors separated by an insulator, which can separate charges under the influence of a voltage, sending electrons to one side and "holes" to the other. This separation of charges, like the separation of chemical ions in a battery, stores energy. The first capacitor was the Leyden jar, invented in Europe in 1745, which consisted of two sheets of metal foil: one on the inside of a glass jar, and one on the outside. The device was considered a novelty, as it stored enough energy to deliver a painful shock to anyone who touched its electrode.

Commercial "ultracapacitors" today consist of carbon electrodes riddled with microscopic pores. These pores don't affect the capacitance of the material, but they do vastly increase its surface area, which is where the electrons are stored. Ultracapacitors are quite dangerous because their storage capacity is comparable to that of batteries, but they can be charged or discharged almost instantaneously. This presents an electrocution hazard to the unwary, but also makes capacitors much better than batteries for high-demand applications like electric automobiles. So ultracapacitors have a bright future, and replacing their "holey carbon" with quantum dot materials may allow the storage of even higher numbers of electrons. Storing electrons is, after all, what quantum dots are all about.

Still another type of storage device, the superconducting battery, simply consists of a loop of superconducting wire with current circling through it endlessly. These devices are 100 percent efficient, but also difficult to operate and maintain with today's technology—the loop has to be immersed in liquid helium at 3°K in order to function. Possible?

Absolutely. Practical for home use? Not really. But we've already looked at the possibilities of high-temperature superconductors, and if these pan out, then quantum dot materials may provide an even better means of storing energy.

Heat Flows

Another form of energy is heat, which by its nature flows out of hot places and into cold ones, in the same way that water runs off of mountaintops and into depressions in the ground. "Radiant heat" can pass through air or vacuum in the form of infrared light waves, and can be "convected" from one place to another by currents of gas or liquid. But heat can also travel directly through matter, in the same way that electricity can. In fact, materials that make good electrical conductors typically also make good thermal conductors, because in solid and liquid materials, electrons are the main transporters of energy.

As we've already noted, the electron mobility of a material is related to how full its outer electron shell is. The best way to visualize this is with an analogy: imagine the atom as a tree, and its energy levels as branches at different heights. Now imagine that the electrons are monkeys. Monkeys are energetic and playful—they like nothing better than to leap from tree to tree. Unfortunately, they are also lazy; they will happily climb or leap downward, but will climb upward only if they can't find a place to sit, or if they've been excited—perhaps by being slapped or bitten by another monkey. (See Figure 4.4.)

The "monkey mobility" of a particular forest is a function of how full the branches are. Since the monkeys are lazy, they tend to migrate downward, so the lower branches of every tree are completely filled with monkeys. These monkeys can't go anywhere because the branches around them are just as full, and so are the branches on neighboring trees. In the Insulator Forest, the middle branches are also full, so none of the monkeys can move unless they're excited up into the upper

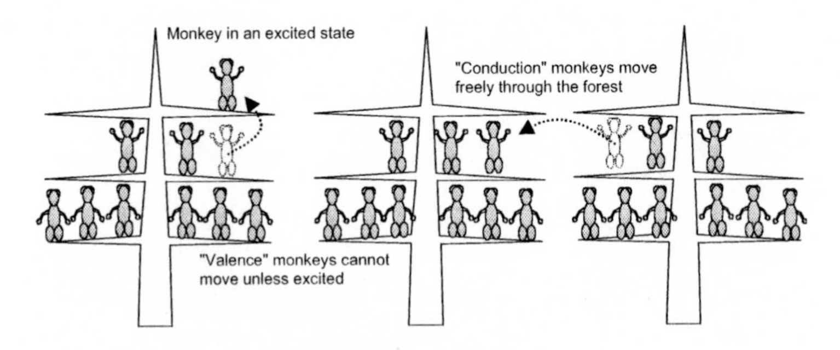

FIGURE 4.4 "MONKEY MOBILITY" IN THE QUANTUM FOREST

Imagining atoms as trees and electrons as monkeys provides an illustration of electrical conductance: the monkeys are free to move only in the higher, emptier branches.

branches. However, the climbing distance to those branches—the band gap—may be large. In reality, there *are* no such things as insulators, just semiconductors with extremely wide band gaps.

In a normal semiconductor, the middle branches are not completely filled, and the band gap is relatively small. With the addition of energy—say, lighting a small fire under the trees—the monkeys can easily be driven up into their conduction band. And in a conductor, the middle branches are less than half-full, leaving plenty of room for monkeys to move around, even without a source of excitation.

What does this have to do with heat? In a gas, heat is conducted by the collision of entire atoms, which smack into each other like billiard balls, transferring energy and momentum. In a solid or liquid, the atoms don't have room to move around, so this energy transfer is accomplished through the electrons instead. Clearly, materials with high electron mobility will tend to be good heat conductors, while materials with low mobility will tend not to. Unfortunately, the situation is a bit more complicated than this. Heat is actually transmitted by "phonons," tiny packets of mechanical vibration that can be treated as either waves or particles. Particles of heat, which travel through a tangle of electrons?

TABLE 4.1 THERMAL CONDUCTIVITY OF SELECTED MATERIALS

Material	Thermal conductivity
Diamond	2000 W/m°K
Nippon Graphite Fiber YS95A	610 W/m°K
Silver	420 W/m°K
Copper	390 W/m°K
Aluminum	230 W/m°K
Silicon	125 W/m°K
Iron	70 W/m°K
Stainless Steel	12-35 W/m°K
Glass	<4 W/m°K
Quartz	1.4 W/m°K
Water	0.6 W/m°K
Wood	0.5 W/m°K
Epoxy Fiberglass	0.04 W/m°K
Air	0.03 W/m°K
Vacuum-Filled Glass Panel	0.003 W/m°K

Yeah, it's confusing. The distinction matters only because there are a number of substances that are excellent thermal conductors, yet poor electrical ones. Diamond is probably the most notable example (see Table 4.1). Others include boron nitride, aluminum nitride, beryllium oxide, and magnesium oxide. This property is associated with "metalloid" elements—atoms near the metal/nonmetal boundary on the periodic table. Metalloids have electronic structures halfway between metals and nonmetals—their outermost electron shells are about halfway filled. This trait is also associated with *light* elements, whose outermost electron shells are small and can't hold many electrons anyway. Marcus adds, "I think high thermal conductivity in these materials is associated with the effect of high stiffness on their phonon spectrum."

Even fewer materials display the opposite property: being good electrical conductors and poor thermal ones. One of these is household dust, a notorious hazard for delicate electronics. Consisting mostly of empty space, "dust bunnies" don't provide much contact area for the transfer of thermal energy. But they do consist mainly of organic molecules, some of them highly polarized, and they readily acquire a static charge. These traits are probably important in creating electron pathways through which small but potentially harmful currents can flow. Also, perhaps more significantly, a number of high-temperature *super-conductors* are now known, based mainly on copper oxides and involving heavy metals like barium and rare earth elements such as yttrium. These materials are ceramics and make excellent thermal insulators. The mechanism for their current-carrying ability may be similar: numerous narrow pathways along which electrons can flow.

Anyway, regardless of how or why a thermal conductor conducts, if a block of it is heated so that one face grows warmer than the other, that heat will quickly flow across to the other face in an attempt to equalize the temperature. Remember: heat flows from hot to cold. The rate of heat flow in a block of material is proportional to the conductivity of the material (k), the cross-sectional area of the block (A), and the temperature difference between the two faces ($T_1 - T_2$). It is inversely proportional to the block's width (w).

The units of "k" are watts per meter per degree Kelvin. This is rather an awkward measurement, difficult to visualize or make sense of, but a comparison of the conductivities of various familiar materials may be instructive (again, see Table 4.1). Notably, the champion heat conductors are various forms of carbon.

In general, the thermal conductivity of a material is treated as a constant. But is it always? Nearly everyone is familiar with electrical switches, but it turns out that thermal switches exist as well, which can turn the flow of heat through a material on or off . Researchers in Bavaria have created electrically switched insulation panels, which rely on the chemical absorption or release of hydrogen gas from a metal-hydride material between two glass panes. When the panel is full of gas—the "on" state—it transmits heat about as well as wood. When the gas is absorbed, the insulation behaves as a vacuum panel, transmitting only a fiftieth as much heat as before. Intermediate states are also possible, although from the standpoint of builders and homeowners, any value in this range will be considered an insulator rather than a conductor. Of course, such thermal switches can easily break or leak, rendering them useless. So it's not a very impressively useful technology, but it does demonstrate an important principle.

We've seen that the addition of dopant atoms to a semiconductor—even in quantities as low as one dopant atom per million atoms of substrate—can dramatically alter the material's electron mobility. We've also seen that electrostatic quantum dots can serve as programmable dopants, with major effects on the electrical conductivity of the underlying substrates. In a quantum dot world, electrical conductivity and resistance are real-time tunable properties, so it's rather a small step to imagine a similar effect on thermal conductivity. "The thermal properties would be completely different from the original semiconductor," Kastner agrees.

Current experiments are already pointing the way. "There's been some very recent and exciting work by Biercuk and Johnson," Marcus says, "on nanotubes added to epoxy, with *very* high thermal conductivity." My ears prick up at this. A material that's as sticky and electrically insulative as epoxy but as thermally conductive as graphite would be a boon indeed to the electronics industry, channeling heat away from chips and other components, allowing them to operate at much higher ambient temperatures.

"By the way," Marcus adds, "Johnson is a good friend of mine, who was a postdoc at Delft, and worked with Kouwenhoven, who was McEuen's postdoc. And Biercuk was an undergrad working with Johnson who is now working on nanotube quantum dots with me. Ours is a small world, no pun intended."

I doubt that quantum dots will lead to materials more thermally conductive than diamond or carbon nanotubes, which may very well represent a theoretical maximum. Still, we're very likely to see "thermal switch" materials that *approach* the conductivity of carbon, yet can be made as insulative as glass, or maybe even fiberglass, when electrically tickled. In fact, there's nothing prohibiting a "thermal rheostat" with numerous in-between settings as well. The industrial applications for such a material are virtually limitless, but the most obvious use is insulating buildings. Given the rising cost of energy, the material could be economical here even if it were fairly expensive. Could there even be thermal switches that are always electrically conductive, or always electrically insulative? The rare materials cited above provide our first clue, and in fact by late 2001, scientists working with layered semiconductors—some with features only a few atoms thick—had made remarkable progress in manipulating thermal conductivity without affecting electrical properties. So the area looks very promising indeed.

Pumping Heat

Heat flow usually happens via passive conduction, but heat can also be *driven* into or away from particular areas. With the input of energy, it can even be forced to defy nature, and flow from a cold area to a hot one. This is exactly what happens in mechanical heat pumps, such as you find in your refrigerator and air conditioner. And it turns out that the same effect can be achieved in the solid state, with no moving parts except electrons.

In 1834, Jean-Charles Peltier discovered that dissimilar metals, when sandwiched together in certain ways, could serve as a thermoelectric generator. When one face of the sandwich was exposed to a heat source and the other to a heat sink, an electrical voltage would be created across the device. The same effect works in reverse, as a heat pump: when a current is passed through the sandwich, one face grows hot while the other grows cold. (See Figure 4.5.) It was later discovered that the effect was even more extreme when dissimilar semiconductors were used.

A mature technology since the 1960s, semiconductor Peltier junctions work very well, and if their current draw is not limited with a resistor, one face can easily burn up even as the other is collecting ice from the atmosphere. This inexpensive technology is often used to cool circuit boards—especially the imaging elements of charge coupled device (CCD) cameras, although you do need a heat sink or thermal shunt to carry the heat away. Many "green" refrigerators today employ this tech-

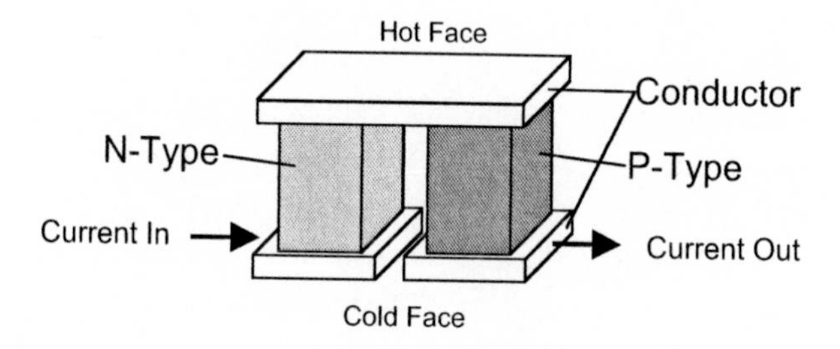

FIGURE 4.5 PELTIER JUNCTION AS HEAT PUMP

Running an electrical current through this device will drive phonons (thermal energy) into the upper conductor, which grows hot. The lower conductors lose energy and become cold. The same principle works in reverse: heating the top conductor (or cooling the bottom one) will create an electrical voltage across the device.

nology, although with present materials they are less efficient than mechanical models, and can only drop the temperature to about 40 degrees below ambient (and raise it to 40° above at the heat sink), so their use is generally restricted to portable refrigerators that can be plugged into the electrical system of a boat, car, camper, or hotel room. For these uses they are advantageous because they're slim, cheap, lightweight, and solid-state.

There is a theoretical limit to the efficiency of any heat pump, regardless of design. Even with perfect materials behaving perfectly, the maximum possible efficiency is 100 percent when the hot and cold faces are at the same temperature, and around 50 percent when the cold face is at half the absolute temperature of the hot one. The rest of the energy is lost as waste heat. And the materials in today's Peltier junctions are very far from perfect, thanks to a tradeoff between electrical resistivity and heat conduction. Natural materials offer no good solution; the best efficiencies achieved with them so far are around 10 percent. An optimal thermoelectric material would be simultaneously a strong electrical conductor—perhaps even a superconductor—and a smotheringly good thermal insulator.

Fortunately, an approximation of this ideal material actually exists, at least in the laboratory. In 2001, scientists at the Research Triangle Institute (RTI) in North Carolina used atomically precise, layered semiconductors—similar to quantum wells—to create a Peltier junction that operates at 2.5 times the efficiency and 23,000 times the speed of all previous models. The electrically semiconductive materials were unusually good radiators—and unusually poor conductors—of heat. The fabrication process was lengthy and expensive, but also similar in many ways to the market-proven production of nanolayered read/write heads for hard-disk drives. And the researchers are predicting another performance doubling (or more-than-doubling) in the near future. Says RTI's Rama Venkatasubriaman, "We've made a pretty good P-type material. We still have to improve the N-type, and the device design, and a lot of manufacturing issues. This is where the big payoff will be." So in coming years, we'll almost certainly see commercial products based on improved forms of this technology.

In this discussion, it's hard not to notice the clustering of important research on America's east coast. Could it be the proximity to European sites such as Delft? "Perhaps," Marcus allows, "or perhaps it's the 'east coast' personality or that Stanford is more engineering oriented, or that the weather in California is too nice for hard work. The fact is that there are isolated pockets of excellence on the west coast. Santa Barbara is a great school for this kind of stuff, for instance. The move east is kind of new. Recall that four years ago, McEuen and I were out west. There's also Roukes at Caltech, who is excellent. So it may just be that the density of schools is higher in the east."

Regardless of where they're made, the most immediate application for improved Peltier junctions will be in ultraquiet refrigerators and air conditioners. Since the technology is solid-state, it can be fashioned into a variety of shapes and probably even made flexible. Peltier "cool suits" may someday harness solar power to pump heat away from the wearer's skin (and blast it at unwary passersby). There may even be portable folding ice rinks, in the same way that today there are inflatable swimming pools. Just remember not to touch the heat sink, which could easily double as a barbecue grill!

Flipped over, these super–Peltier junctions could also serve as thermoelectric generators, capable of capturing waste heat from many devices or processes and converting it back into electricity. This is already done on board deep-space probes such as Galileo (visiting Jupiter) and Cassini (visiting Saturn), which harvest the heat from radioactive plutonium in a device known as a radioisotope-thermal generator or RTG. Despite public hysteria, RTGs are in fact so rugged that they've not only survived the explosion of launch vehicles, but have been recovered from the ocean floor afterward, and re-flown on other missions. Closer to home, super–Peltier junctions could recapture the heat of car engines, or even the heat of quantum dot devices shedding large numbers of electrons.

Squeezably Soft

Still another way for quantum dots to harvest energy is through the "piezoelectric effect." This occurs when pressure on a material creates slight dipoles within it, by deforming neutral molecules or particles so that they become positively charged on one side and negatively charged on the other, which in turn creates an electrical voltage across the material. This is the operating principle behind an old-fashioned phonograph: the needle, moving over the surface of a vinyl record, encounters ridges and grooves that deform its Rochelle-salt tip, creating voltage patterns that are amplified and converted into sound. Interestingly, the production of sound can also rely on the piezoelectric effect, or rather on its reverse: a voltage applied across certain materials will cause them to deform, and an oscillating voltage will make them vibrate. These vibrations are then acoustically focused and amplified, in the same way that guitar-string vibrations are amplified by a hollow box of wood. This is how ceramic stereo speakers work.

Zinc oxide, the most piezoelectric material presently known, is widely used in electronic transducers, or solid-state devices for converting electricity into mechanical force. It has other strange properties as well—it's one of the whitest pigments available (although titanium oxide is whiter). When doped with aluminum it's also a semiconductor, and when doped with barium or chromium oxide it has the most non-Ohmic electrical properties of any known material. Ohm's Law, which holds true for virtually every substance on Earth, states that electrical voltage is equal to current times resistance, or $V=IR$. In these zinc oxides, though, the equation becomes $V^n = IR$, where n can be upwards of 100. You get huge amounts of voltage for free, without paying the usual penalties in high current and resistive heat losses. This bizarre effect is used in electronic components called "varistors," which are commonly used as surge arrestors in noisy electronic circuits. Feed in a huge voltage spike and the current varies only slightly, preventing damage to any delicate electronics downstream.

If it were more common and more controllable, this effect could serve as the basis for entirely new components and entirely new types of electrical circuitry. And since it is associated with piezoelectricity, any progress in this field would also be progress in the development of new piezoelectric devices.

"Look around," Marcus says, "at the spectrum of materials that exist in the world, with all their various thermal and electrical properties. These are all made from a handful of naturally occurring substances. By constructing materials out of man-made objects instead, we can introduce a new subtlety. The first step is putting one material next to another—it's amazing what you can accomplish that way. Imagine what you can do when the fundamental building blocks are *designed* to assign a personality to macroscopic materials. That spectrum of possibilities gets a lot broader."

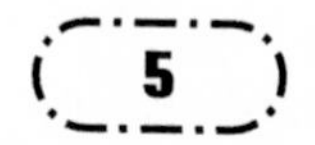

Magnetism and Mechanics

Material objects are of two kinds, atoms and compounds of atoms. The atoms themselves cannot be swamped by any force, for they are preserved indefinitely by their absolute solidity.

—Titus Lucretius Carus, *On the Nature of Things*, circa 50 B.C.

QUANTUM DOTS HAVE still other capabilities, but to discuss them requires some understanding of what the electrons are doing when they're trapped inside. Nobody knows more about this than MIT professor Raymond C. Ashoori. The trip back to MIT is not a difficult one; just an eastward jaunt down Massachusetts Avenue, toward the wide spot in the Charles River. Follow the smell of water, and swerve to the left at the last opportunity, or risk crossing over once again into the wilds of Boston. Maneuvering into a parking garage is somewhat more involved, since a lot of the streets are one way, face the wrong direction, and spike off in the weird angles characteristic of older cities. Even so, it doesn't take long to reach the Infinite Corridor, and the Research Laboratory of Electronics, and finally the domain of Ashoori himself: the Quantum Effect Devices Lab.

The digs are fairly humble compared to those of Kastner and Marcus—the room looks more like an ordinary professor's office at an ordinary college. There's also nothing flamboyant about Ashoori's own appearance—on the street he could be just another late thirtysomething, slowly losing the hairline battle. Nor does he draw attention to himself with dynamic vocals or gestures. He is the proverbial quiet man.

But looks, as they say, can be deceiving. Ashoori is another Bell Labs alumnus, and since landing at MIT in 1993 he has had cover stories in *Nature, Science,* and other top journals, and has won a half-million-dollar fellowship from the David and Lucile Packard Foundation, over and above the grants and salary that keep his research moving. He has also won a Sloan Fellowship, a MacMillan Award, and the National Science Foundation's Young Investigator Award. He speaks five languages fluently. He draws and paints. He keeps in shape. At any other school, he'd be a superstar of the highest order, but this is MIT, where mere brilliance is the minimum qualification.

Ashoori became known early on for constructing an artificial atom that could hold anywhere from zero to fifty electrons, but his truly groundbreaking work has occurred in the field of imaging. Looking at very small things is always a challenge—even more so when they're smaller than a wavelength of light. Traditionally, an electron beam is used to image mesoscopic objects, and can even capture features as small as 1 nm. An electrically charged atomic force microscope (AFM) tip goes one better; it's actually capable of seeing (well, feeling) the atoms these objects are made of. But when you want to look at *individual electrons*, you've got problems.

In 1991, Ashoori devised a technique called "single-electron capacitance spectroscopy," which could image the contours of electron density (or probability density) in the "2D electron gas" of a quantum well or large quantum dot. This technique treats charge as a poorly localized cloud and is well suited to the quantum, wave-like behavior of confined electrons. Charge contours show up in the image as abrupt color changes, so that an area containing (on average) three electrons is brighter than a surrounding area that contains only two. It's exactly like looking at a topographical map.

Another technique—AFM Quantum Transport, pioneered by Harvard's Robert Westervelt—is closely related to Ashoori's, and relies on placing an obstacle (the probe tip of an AFM) in the path of moving electrons and measuring the resulting charge distributions. By taking endless measurements with the obstacle in slightly different positions each time, researchers can patiently deduce the classical trajectories of the electrons, treated as discrete point-charges or particles. If this is difficult to visualize, go back to the hockey-rink metaphor: imagine we have a cannon that shoots out pucks from a particular location and in a particular direction. We also have a scanning technique that maps out the statistical distribution of the pucks. Areas where a puck frequently crosses are brighter, and those less frequently touched are darker.

If successive images are taken while a Zamboni machine (representing the probe tip) is moved around on the ice, then the deterministic, repeatable path of the puck can be computed by studying the way it bounces off the Zamboni in different locations. The resulting image is not a cloud at all but a narrow line—a trajectory. If Ashoori's technique gives a contour map, then Westervelt's yields a road map, showing exactly where the electrons are and are not driving as they cross the terrain of the quantum well. These paths, which look remarkably like lightning bolts, are precisely the classical complement to Ashoori's quantum imagery, and together the two techniques are beginning to show us how quantum dots really behave inside. (See Figure 5.1.)

A further wrinkle comes when Ashoori applies a high magnetic field to the electron gas. As the field strength changes, the energy levels of the electrons shift up and down—the monkeys on their tree hopping to higher and lower branches. Measuring how much they move is important, because the response of the artificial atom to the magnetic field reveals an important detail about the individual electrons: their spin. This parameter is intimately related to the physical shape of electron shells, which are actually made up of smaller units called "orbitals." The exact structure of these orbitals—the waveform of the confined electrons—is critical in defining the magnetic and chemical properties of the atom.

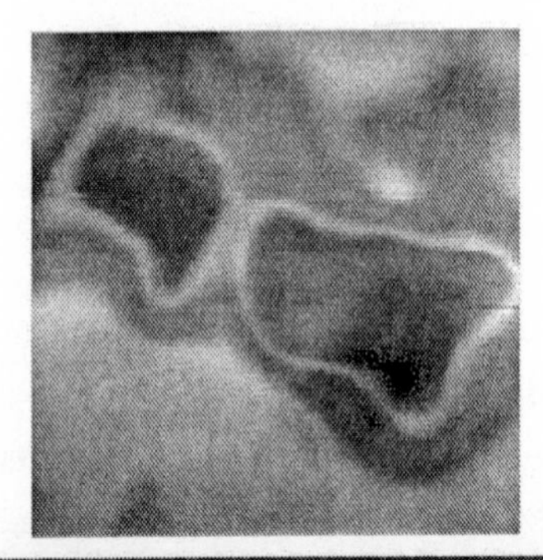

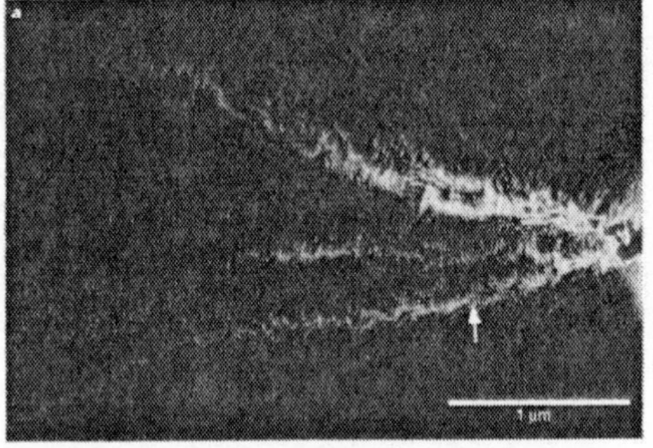

FIGURE 5.1 CAPACITANCE SPECTROSCOPY AND QUANTUM TRANSPORT IMAGES OF A TWO-DIMENSIONAL ELECTRON GAS

The image on top (courtesy of Raymond Ashoori) is a contour map showing the quantum (wave-like) distribution of electrons in a quantum well. The image below it (courtesy of Robert Westervelt and Eric Heller) shows the classical (particle-like) path of electrons passing through a similar well. As these techniques are refined to scan smaller areas, they may eventually image the electron orbitals of an artificial atom.

Electron Structure of Natural Atoms

This is where our monkey analogy starts to break down. To this point, we've considered only one significant parameter: each monkey's energy level, or which branch it's sitting on. In quantum mechanics, this is known as the "principal quantum number," and it is designated by the letter "n." This number is extremely important, but by itself it does not complete the description of electrons in an atom. Electrons have three other parameters of critical importance: angular momentum ("l"), magnetic quantum ("m"), and spin ("s"). There is another behavioral quirk of electrons that we now need to know about: the Pauli Exclusion Principle. This rule states that no two electrons in an atom can have the same quantum numbers. In monkey terms, we would say not only that every monkey is unique but that monkeys actually repel one another, and will sit as far apart as their laziness permits. One result is that as the principal quantum number increases (as we move further away from the atom's center), we have more space available, so the number of electrons each shell can hold also increases.

Rather than belabor the theory any further, I'll simply describe the shells of a natural atom. The two lowest are the 1s and the 2s, which are spherical and hold two electrons each. The next three are the 2p orbitals, which are dumbbell shaped and fit into the atom along perpendicular axes. A total of six electrons can fit here. Applying these rules to a simple atom, we can see that helium should be small and perfectly spherical, and should also be an insulator, since its outer electron shell—its only electron shell—contains two electrons, the maximum it can hold. As we'll see, this configuration also makes the helium atom nonmagnetic and chemically inert.

Carbon is slightly more complex: the atom has its 1s and 2s orbitals completely filled, with two additional electrons slotted into the 2p orbitals. (See Figure 5.2.) As you might expect, the carbon atom is larger than helium, and has the dumbbell ends of the p orbitals sticking out of it. Since the outer shell—level 2—can hold a total of eight electrons, it's a little less than half-full. Carbon is therefore a metalloid and can behave as a conductor, semiconductor, or insulator depending on how the atom is bonded to its neighbors. Carbon is also weakly magnetic and chemically reactive. In fact, carbon can bond with so many different things in so many different ways that an entire field of study—organic chemistry—is devoted to it.

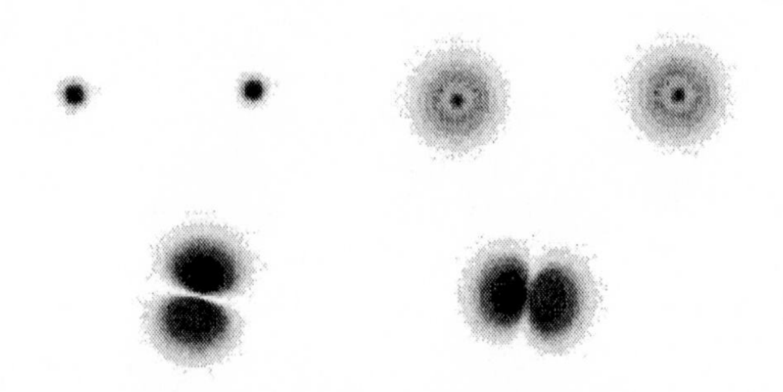

FIGURE 5.2 ELECTRON ORBITALS OF THE CARBON ATOM

The 1s orbital is a tiny sphere containing two electrons. The larger 2s sphere surrounds this with two electrons of its own. Two dumbbell-shaped 2p orbitals contain one electron each, giving the carbon atom its characteristic pyramid or clover-leaf shape. The "arms" of the 2p orbital can flex dramatically as the carbon atom bonds with neighbors in various ways.

Level 3 has one s orbital and three p's—just like the ones at level 2, except that they're twice the size. At this level we also encounter a new set of orbitals: the 3d's. There are five of these, shaped for the most part like four-leaf clovers. At level 4 we find a 4s, three 4p's, five 4d's, and the most capacious* of the natural orbitals: the 4f. This holds fourteen electrons and looks vaguely like one of those "onion blossom" appetizers you can get at trendy restaurants. Because its electrons are less strongly bound than the ones closer to the nucleus, the 4f orbital is easily excited and can exhibit thousands of distinct energy states. It is associated with the unusual optical and magnetic properties of the rare-earth elements. (See Figure 5.3.)

Why is this orbital structure important? All of chemistry is based on it, for one thing, and it's also critical to the phenomenon of magnetism.

*The 5s, 5p, 5d, 6s, 6p, and 7s orbitals are physically larger but contain fewer electrons.

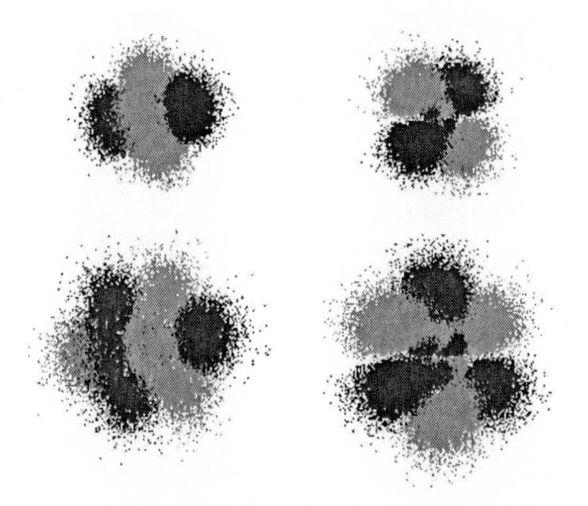

FIGURE 5.3 MORE COMPLEX ORBITALS

These d (top) and f (bottom) orbitals are more complex than the s and p ones, and exhibit a broader variety of energy and spin states.

Each complete orbital in the atom consists of a pair of electrons spinning in opposite directions (known for convenience as "spin up" and "spin down"). These spins cancel each other out, whereas orbitals that aren't completely filled (i.e., that contain only a single electron) are "spin polarized" and are attracted to magnetic fields. There are seven electron shells in all, and the dynamics of their filling-up are complex and somewhat counterintuitive. Interested readers should grab a freshman chemistry textbook for a detailed explanation.

Electrons and Magnetism

The important point is that most elements have at least one or two unpaired electrons and are "weakly paramagnetic." We don't normally consider metals like aluminum to be magnetic, but a strong enough magnet will indeed attract them. Some atoms have several unpaired electrons and are much more strongly paramagnetic. In iron, for example, five out of twenty-six electrons are unpaired, leaving the atom with a substantial net spin. The electron shells literally whirl around the nucleus, forming a subnanometer-sized dynamo that creates a fixed magnetic field around the atom. Other strongly paramagnetic atoms include palladium, platinum, and the rare-earth elements.

The related phenomenon of "ferromagnetism" occurs when the atoms in a material align so that their spins are all (or mostly) in the same direction. This creates a single, fixed magnetic field around the entire material, as in a natural lodestone or man-made iron magnet. Ferromagnets can attract or repel each other, and can attract paramagnetic materials. Not all paramagnetic elements are also ferromagnetic; the phenomenon is observed mainly in iron, cobalt, nickel, and some of the rare-earth elements, although it can also occur in alloys containing these elements, and in some compounds that contain other paramagnetic elements. The most powerful ferromagnets are iron-neodymium-boron (NdFeB) or samarium-cobalt (SmCo) alloys. One cubic centimeter of these materials can generate an attractive or repulsive force of around 280 pounds—enough to lift a professional football player.

Most ferromagnetic materials are electrical conductors, although a few are semiconductors or insulators. There are also antiferromagnetic materials, whose atoms align in spin-up/spin-down pairs to neutralize any net magnetic field even if the individual atoms are strongly magnetic. In fact, many materials can be either ferromagnetic, antiferromagnetic, or paramagnetic (disordered) depending on their temperature.

The next question is, Can quantum dots exhibit this property as well? The answer is: yeah, probably. Experimentally, it's very difficult to measure the spins of the electrons in a quantum dot, and even more difficult to map the orbital structure. This difficulty increases as the quantum dot grows smaller and more atom-like. Also, in the theoretical world it's very difficult to model the behavior of large numbers of confined electrons; the number of calculations increases exponentially with the number of electrons, and anything more than about nine is beyond the capacity of today's computers. This uncertainty means there are no guarantees of magnetic behavior, however likely it seems. Magnetism is the quantum dot's final frontier, the least understood of its potential properties.

We do know that in small, spherical, highly symmetric quantum dots, theory and experiment agree that the first nine electrons arrange themselves much as they would in a natural atom. For pseudofluorine, an artificial atom with nine electrons, there is something like a 1s orbital, surrounded by something like a 2s orbital, with the Q-tip lobes of the 2p orbitals sticking out. As of this writing, no one has pushed this to number ten, but there's no reason to believe that adding one more electron to create pseudoneon would violate this basic structure. Most likely, the p orbital would be completely filled, turning the atom into a nonmagnetic electrical insulator.

In larger atoms, though, there is a third electron shell, where all bets are off. All we know so far is that, in some instances, the electrons don't occupy the atom in spin-up/spin-down pairs—most especially if the atom is large or asymmetrical. This behavior is complex and not well understood. Furthermore, any deviation from spherical symmetry will radically distort the shape and properties of the orbitals. The most extreme example is a perfectly square or cubical quantum dot, in which the "orbitals" are linear tracks along which the electrons bounce back and forth. (See Figure 5.4.) An electron traveling along the left-right axis would have no way to transition to the forward-backward axis, so the orbital structure would consist of two perpendicular sine waves. If one axis were much longer than the other, any light emitted by the dot would be highly polarized.

In practice, it would be extremely difficult to build quantum dots that were perfectly spherical *or* perfectly cubical. Real quantum dots will be only approximately spherical, or perhaps cylindrical, pyramidal, or ellipsoidal. They will vary slightly in size and shape and will likely include

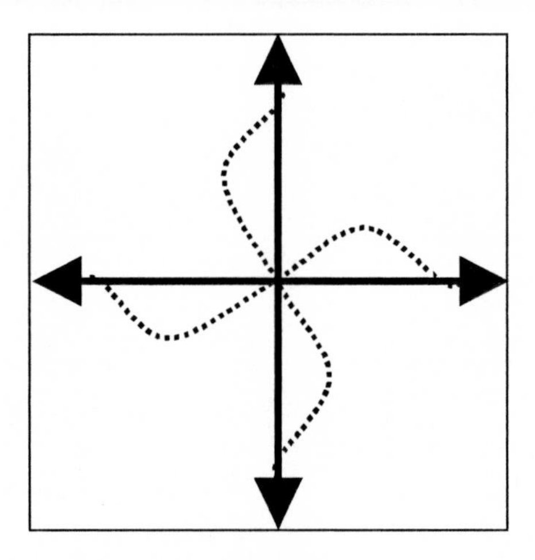

FIGURE 5.4 SQUARE ATOM

The "orbitals" of a square atom consist of perpendicular sine waves.

some misplaced dopant atoms or other impurities, which throw off their symmetry still further. The concepts of orbital chemistry date back to 1870, when Dmitri Mendeleev published the first periodic table, and were heavily refined in the twentieth century based on the discovery of quantum mechanics and specifically on Erwin Schroedinger's wave equations for the electron. They are adequate for describing the specific orbitals found in natural atoms, but they require a major update for the description of the more general case, where electrons are confined by large, nonspherical, non-nuclear fields or structures.

This, of course, is where experimentalists like Ray Ashoori are pinning their dreams. Any theory or model is based on a hypothesis, and every hypothesis is based on observation. When enough observations pile up, the new theory may suddenly become obvious to the people standing in the right place. Either that, or unimaginably powerful new computers will make possible the brute-force computation of new orbital shapes from the raw quantum field equations. This seems less likely, though, when you realize just how many electrons a quantum dot can hold.

One problem with natural atoms is that nuclear forces limit the number of protons in a reliable atomic nucleus to 92, the atomic number of uranium. Nuclei larger than this (also known as "transuranic" nuclei) contain loosely bound particles that fly out randomly, in a process we know as radioactivity. Some transuranic atoms have long half-lives, such as plutonium-242 (^{242}Pu) at 380,000 years—another way of saying that this particular isotope is not very radioactive. Other forms of plutonium are shorter-lived, such as ^{240}Pu at only 6,580 years. In general, the farther you go past uranium on the periodic table, the less stable the nuclei become. Americium-243 (with 95 protons) has a half-life of 7,370 years, whereas the longest-lived isotope of Lawrencium (103 protons) lasts only 35 seconds. Atoms larger than Lawrencium typically do not appear on the periodic table, since their lifetimes are tiny fractions of a second, so that their chemical and physical properties are largely unknown.

Physicists predict a "stability island" around atomic numbers 114–120, where a handful of elements may be found with half-lives in the hundreds or thousands of years, but this has yet to be demonstrated experimentally. Even if it turns out to be true, it doesn't mean we can make practical use of the optical, electrical, thermal, magnetic, or chemical properties of these atoms—although these will surely be interesting, given the large number of electrons and the complex structures of their outer shells. Since protons and electrons are paired together in the atom, natural atoms containing more than 92 electrons are always radioactive. So except in applications where radioactivity is desirable (such as medicine, weapons, and power generators), these elements won't be good for much.

This is where another "killer app" for artificial atoms comes in: since they're not burdened with a nucleus, they can remain stable with hundreds or even thousands of electrons crammed inside, forming gigantic new orbitals that classical chemists could never have imagined. It's one of the real hinterlands of turn-of-the-millennium physics: the study of highly transuranic artificial elements. The computing power required to model these elements is well beyond the capacity of the entire earth circa 2001 (1 EM in the terminology of Chapter 2), and the problem is unlikely to become tractable anytime soon. But with quantum dots, it becomes possible to study these elements experimentally. Ashoori, though a leader in the field, does have company, most notably Paul L. McEuen of Cornell University, and also to some extent Charles Marcus and Marc Kastner, plus a few others whose work is less familiar but perhaps equally groundbreaking.

Their finding: much weirdness. First, the orbitals don't seem necessarily to radiate outward from a single center the way they would in a natural atom. The orbitals also obey different energy rules—some even exhibit "electron bunching," a phenomenon that permits one or more electrons to slip in and out of the atom with zero energy cost. (See Figure 5.5.) In some cases, up to six electrons can be added for free in this way. Ashoori has found that this occurs when the artificial atom splits into two pieces: an inner circle and an outer ring, or perhaps a pattern of multiple disconnected spots—something that would never happen in a natural atom. To date, this has been observed only in very large artificial atoms containing hundreds to thousands of electrons. Whether something similar occurs in smaller transuranics is unknown.

However spectacular the images are, Ashoori seems to regard them as a kind of intermediate step between detection methods and theoretical analysis. Physicists, being hackers of a sort, are nearly as proud of their methods as they are of the results they get. In discussion, the two are so deeply intertwined that they seem to be conceptualized as one thing, and Ashoori is a case in point. He is hacking matter itself, and I get more information while he's showing off his microscope than I do when he's talking me through the pictures it has taken. He also seems sheepish about his lab's dilution refrigerator and the complex hardware

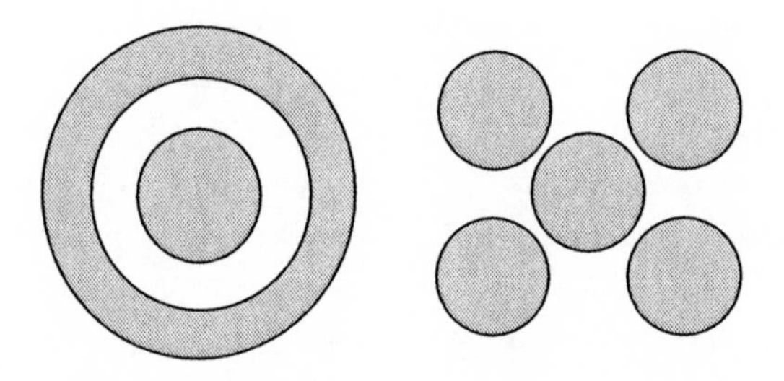

FIGURE 5.5 POSSIBLE ORBITAL CONFIGURATIONS OF A HIGHLY TRANSURANIC ELEMENT

In some atoms, single electrons can be added at no energy cost, implying that the electron cloud has broken into two distinct pieces such as a ring surrounding a central sphere. In other cases, five or more electrons can be added for free, implying an even more complex arrangement. These structures would never occur in a natural atom, and their properties are largely unknown.

that supports it. "If we're going to do anything useful, we're going to have to wean ourselves off these cryogenic temperatures. This field is really in its infancy."

Indeed. The only area that even begins to overlap is the study of metallic nanoparticles, and even here the differences are considerable. Because metals have high electron mobility, metallic quantum dots contain huge numbers of electrons (10^7 is typical), so that their spectrum of energy levels is very crowded and essentially continuous, rather than sparse and discrete as in an atom. It's the difference between a staircase and a ramp. The electrons inside such a dot form an "artificial atom" only by a great stretch of the term. In fact, these particles are something quite different from both bulk metals and ordinary artificial atoms. They are in some sense equivalent to extreme transuranic elements observed at extremely high temperature. They also differ from atoms in that their electrons are confined to the particle's surface, as in a buckyball, whereas semiconductors confine electrons on the inside. Still, metallic quantum dots do provide some useful analogies.

Another finding Ashoori and his colleagues are mulling over is that larger orbitals are much more strongly influenced by magnetic fields than small ones are. Effects can be easily produced in quantum dots that would appear in atoms only at field strengths of a million Tesla or more—about ten thousand times stronger than we can presently create. So artificial atoms will interact with their environment in ways never previously observed. And in larger quantum dots—particularly asymmetric ones with very large numbers of electrons—the shell structure breaks down altogether, yielding a material known simply as "electron gas" whose magnetic properties are anyone's guess.

Magnetic Dots in the Real World

For the moment, magnetic applications for quantum dots are limited, although the field is slowly opening up. "Magnetoresistive" materials developed at IBM involve sandwiched layers of magnetic and nonmagnetic material less than 1 nm thick. Quantum wells, in other words. The resulting materials are about a million times more sensitive to magnetic fields than ordinary materials such as iron. The presence of a magnetic field changes the electrical resistance of the material, which then affects the behavior of an electrical circuit. Although relatively new, these materials are already used in electronic compasses, metal detectors, and especially the read heads of hard-disk drives—the application for which they were originally developed. All of these quantum well devices outperform their classical counterparts by many orders of magnitude and are in high demand in the computer and robotics industries. Dead-reckoning navigation systems employing magnetoresistive compasses, for example, have begun to replace or supplement GPS receivers for some applications; often they're simply more accurate and reliable, especially over short distances and in obstructed areas where satellite signals are difficult to receive.

It's not clear whether quantum dots can improve on this performance, but one thing they should be uniquely able to do is modify their magnetic properties on the fly. This could be accomplished by either pumping electrons in and out, moving the electrons to excited states where their spins are different, or distorting or reshaping the orbital structures to achieve a particular effect. One of the scientists I spoke with boasted privately that his lab would produce a switchable ferromagnetic material by 2006. "I know just how to do it," he mourns, "but I'm a college professor. I write grants and reviews and papers; there's not much time for actual work. What I need to do is find the right grad student to work on it for me."

Meanwhile, IBM is also investigating nanoparticle films, consisting of 3-nanometer quantum dots made of iron and platinum atoms. These should make excellent read/writable magnetic materials for the surfaces of hard disks. In principle, each quantum dot could serve as a bit, which would mean about a hundred-fold increase in storage density over the best materials available in 2002.

Interestingly, nature hopped on this same bandwagon billions of years ago, with the advent of magnetically sensitive bacteria. These tiny creatures contain a few dozen nanoparticles, each between 35 and 120 nanometers in diameter, of an iron-rich mineral called magnetite. This quantum dot-like arrangement results in the maximum magnetic sensitivity for a given amount of iron, because the fields associated with larger particles would break up into multiple conflicting bubbles, while smaller particles would not be able to retain their ferromagnetism for as long. Evolution, the great optimizer, has provided these bacteria with a navigational reference as accurate as a dime-store compass, but using a hundred million times less material.

There are even efforts under way to attach iron-rich nanoparticles to antibodies and other biological molecules, providing the magnetic equivalent of Quantum Dot Corporation's optical markers. When floating in solution, these markers rotate freely and rapidly, so that their magnetic fields smear into an undetectable blur. When stationary (perhaps because the antibody has attached itself to a biological docking site), the fields are strong and easily detected. This approach would map the functions of a living cell in magnetic fields, rather than colored lights. Why does this matter? Mainly because it will permit researchers to view the inner workings of cells they can't directly see, such as tumor cells inside a living person.

And if quantum dots can be coaxed into superconductivity, they may have a final magnetic trick to offer: the Meissner effect. This phenomenon enables some superconductors to expel magnetic fields from their interior, forming effective magnetic barriers. In a popular demonstration, this effect can be used to levitate a wafer of yttrium-barium-copper oxide: just place it on a permanent magnet and pour some liquid nitrogen over it, and the wafer will float up into the air, supported by the pressure of the magnetic field it's reflecting. It's quite stable, with very little tendency to slide off to the side and fall; you can even poke it with a pencil, or spin it. This is a perfect magic trick for parties and school assemblies, because of course it isn't a trick at all but something very close to actual magic.

The Chemistry Set

Artificial atoms, while remarkably similar to natural ones, are clearly capable of a wide range of electronic structures, characteristics, and behaviors that do not occur in nature. In fact, the 92 "natural" structures are tiny and by no means preferred islands in the sea of this technology's overall capability. This we now know: "doping" a semiconductor with artificial atoms can modify its physical, optical, and electrical properties in unnatural ways, with decidedly unnatural results.

What's more, the electrons in two adjacent quantum dots will interact just as they would in two real atoms placed at the equivalent distance, meaning the two dots can share electrons between them, meaning they can form connections that are equivalent to chemical bonds. Not virtual or simulated bonds but real ones—physicists Daniel van der Weide of the University of Delaware and Werner Wegscheider of Munich University have demonstrated that the back-and-forth motion of electrons between bonded quantum dots is strikingly similar to the motion between atoms in a natural molecule—even when the dots are separated by tens of nanometers.

This fact leads naturally to the idea that artificial atoms might be useful in chemistry. If they can interact with each other, sharing electrons between their outer shells just like natural atoms do, then they should also be capable of interacting with natural atoms. That's chemistry, right? Unfortunately, it's not quite that simple. First of all, by definition, the artificial atom is trapped inside a substrate. It's not an object in its own right but a charge discontinuity—a dopant that can't exist outside the matter being doped. Also, artificial atoms are much larger than natural ones, so even if their electrons can be shared, their orbitals will not fit together in the usual ways. The greatest problem, though, has to do with energy.

According to quantum mechanics, the interatomic "motion" of electrons is actually a sort of blurring—the wave function of the electron smears out to encompass both atoms. From a classical standpoint, the electron isn't moving at all but "tunneling" or teleporting rapidly from one point to another. This tunneling acts like a needle pulling a short thread behind it: it binds the two atoms together, in what chemists refer to as a "covalent bond." There is an energy associated with the bond, which corresponds in effect to the number of windings of thread connecting the two atoms. When the atoms are close, the electron shuttles back and forth very rapidly, and the bond is strong. When the atoms are farther apart, the tunneling decreases, and the electron spends less of its time leaping from one atom to the other. Consequently, there are fewer "stitches" between the atoms, and the bond is weaker. Bond strength falls off with the square of the distance, so doubling the separation between two atoms cuts the bond strength between them by a factor of 4. Since artificial atoms are larger than natural ones, they're necessarily spaced farther apart. They therefore have a lot less binding energy.

As if this weren't bad enough, the energy of quantum dots is also affected by the material they're made of. To put it simply, the binding energy between two quantum dots can't be any stronger than the binding energy between the natural atoms from which they're made. In fact, it's often as little as 10–20 percent of that strength. Lego bricks make a good analogy here: they "bond" with little plastic pegs fitted tightly into cylindrical holes, so that the peg is effectively "shared" between the two bricks. But these bonds pull apart under tension, and any structure you build out of Legos, regardless of its design, will be limited to the tensile strength of these bonds. Connecting two giant Lego structures together will not change this; neither will connecting a big one to a small one, or any other combination. Big, complex Lego structures are generally weaker than small and simple ones. This doesn't make quantum dot chemistry impossible, but does restrict it severely, to a domain that Sun Microsystems' Howard Davidson refers to as "the chemistry of very low energies."

In nature there are two basic types of chemical bonds: the covalent bond, which arises from the physical sharing of electrons between atoms, and electrostatic bonds, which rely on differences in charge between two atoms or molecules. The energy of a covalent bond—that is, the energy input you would need in order to break the bond—falls in the range of 1.25–12.5 eV, depending on the atoms and their spacing. But double and even triple covalent bonds can form when multiple electrons are shared between the same two atoms. Diamond, one of the hardest naturally occurring materials, consists of carbon atoms linked by four 5eV bonds each. Covalent bonds literally hold the universe together; they're responsible for most of the strength and hardness we observe in most materials, both organic and inorganic.

The major exceptions are metals and salts, which are held together with much weaker electrostatic bonds. The three main types of these are pi bonds (0.25–0.3 eV), hydrogen bonds (0.04–0.3 eV), and ionic bonds (around 0.06 eV). Some materials, such as water, have both a covalent and an electrostatic nature, making their physical properties complex and interesting. This is why water makes such a good solvent, and also why it expands when it freezes—something few other materials do.

Imagine for a moment that we have a 1.4 nm buckyball: a hollow sphere made up of sixty carbon atoms. It's extremely inert, since every exposed electron in it is shared between two atoms. Chemically speaking, the molecule is blank, invisible, nonreactive. However, it can and does conduct electricity along its exterior surface, and if we trap electrons there it will act as a quantum dot. In fact, if we trap six electrons there it will act in many ways like a single large carbon atom, ten times the diameter of a natural one. The orbitals will be highly distorted, since they're squished onto the surface of a sphere, but in terms of levels and spins they will probably look a lot like real carbon. If another atom should happen along—say, an oxygen—it would find four valence electrons on this pseudocarbon, and four empty spaces where an electron could go.

The electrons are at a lower energy, though, and are less available for binding. The oxygen might very well bind with the buckyball, sharing two or three of its excess electrons, but the resulting molecule would not have the chemical or physical properties of carbon monoxide. The bond would be only a hundredth or even a thousandth as strong—maybe 0.01–0.1 eV—so at room temperature the oxygen could easily be knocked away by the thermal vibrations of other atoms and molecules. Still, at the high end of this range, the bond is stronger than most ionic and hydrogen bonds, and approaches the strength of a weak pi bond.

And remember, artificial atoms don't just sit there: like spreadsheet or automaton cells, they interact with their neighbors. We can not only form chemical bonds between them but also turn the bonds on and off as electrons are pumped in and out. This is chemistry: weak but real. Nano- and microtechnology promise to rearrange the shape and texture of materials, which is great, but not really so different from what we can already achieve manually, with a machine shop or even a simple potter's wheel. Molecular nanotechnology may even be able to rearrange atoms, albeit slowly, like a sort of mechanized plant or fungus. But here is something entirely new: a material capable of *changing its very substance*, instantaneously.

Biological receptors with bond strength of as little as 0.6 eV can reliably attract and hold target molecules 99 percent of the time, and weaker ones exist as well, so it is highly probable that buckyball quantum dots would have observable medical effects in the human body (see Chapter 7). In fact, raw buckyballs have already proven useful in medicine thanks to an even weaker force: the van der Waals bond, which causes neutrally charged particles to be attracted together. Between two small molecules, this bond energy is only 0.025–0.05 eV, although the process favors some materials over others, and also (importantly) scales with the frontal area of the particles, so that the bond between a buckyball and a small organic molecule is higher—more like 0.1 eV.

The van der Waals bond between two C_{60} buckyballs is around 2 eV—quite comparable to a covalent bond—and the energy between two Bawendi-type quantum dots is in the same range. This leads to another obvious question.

Superstrong Materials?

Covalent bonds between quantum dots are weak, but with more electrons to share than natural atoms, and many more ways to share them, it seems that we should be able to form not just double or triple bonds but massively parallel ones. If a hundred electrons were shared, wouldn't that make the bond a hundred times stronger? And with van der Waals forces creating interdot bonds much stronger than *any* covalent bond, natural or otherwise, shouldn't colloidal materials be much stronger than atomic ones? Alas, the answers are no, and no.

Sharing large numbers of electrons means *storing* large numbers of electrons, which means filling up all the low-order electron shells and branching out into vast, low-energy orbitals that are not at all tightly bound. Tug on the electrons with a chemical attraction, and they will simply pull free. They're like broken Lego bricks—they'll stay where you put them, but they can't really grip. Any disturbance will knock them apart. As in nature, the strongest covalent bond you can form is probably a triple sharing of the electrons in the 2p orbital.

Still, that 2eV van der Waals bond looks pretty impressive. Random thermal energy certainly isn't going to break it, so we can guess—correctly—that buckyball crystals should be stable at room temperature and above. But we've already seen Bawendi's colloidal materials, which are held together exclusively by these bonds, and they are by no means superstrong. Quite the reverse, they're extremely brittle and easily scratched, even though the CdSe semiconductor they're made from is rather tough. Where does this discrepancy come from?

Because real atoms are smaller than artificial ones, they're also closer together. Atomic crystals are densely packed, and so are the bonds between the atoms. Diamond is made of 5eV covalent bonds, and a 10 nm sphere of it contains over three thousand of them, oriented in a variety of directions. You wouldn't need to break *all* of these bonds in order to destroy the crystal, but you would need to break all the ones that pass through one particular plane of cleavage. This would tear (or shear) one half of the sphere away from the other half, which is exactly what a diamond cutter does when shaping jewels. How much energy would this take? About 500 electron-volts, give or take. Contrast this with the ~2 eV required to pry apart two spheres of the same size joined by van der Waals forces, and you quickly see why it's easier to separate colloidal dots than it is to break them.

Moreover, since the cross-sectional area of a crystal grows as the square of its diameter, this effect scales enormously with size. If we have a film of Bawendi dots that is 30 nanometers thick and 1 million nanometers (or 1 millimeter) wide, cracking it requires only the breaking of 300,000 van der Waals bonds. Cracking a diamond film of the same dimensions requires the breaking of *billions* of covalent bonds, using thousands of times more energy. This is why diamond films make great, durable coatings for tools, while colloidal films do not. The difference becomes even more enormous when you start talking about bulk crystals.

Still, if Ashoori and his crowd can figure out a few things about electron spin, then the door may not be completely closed on strong quantum dot materials. Rheology is the study of how materials deform, specifically the difference between plastic (i.e., permanent) and elastic (i.e., temporary) deformation. Does a material flow like wax, or is it springy like rubber, brittle like glass, or as firm as granite or basalt? The rheological properties of most solids are strongly related to temperature—rigid when cold, softer and more fluid when heated. With enough heating, there comes a phase change when the solids lose their crystalline structure and become liquids. (Certain organic materials, such as wood and thermoset plastics, do not behave this way, but they're the exceptions.) Interestingly, though, there is a class of materials whose rheological properties are just as strongly affected by electricity and/or magnetic fields.

The simplest magneto-rheological (MR) materials are suspensions of ferromagnetic powder in heavy oils. These materials are normally liquid, but they can be coaxed by powerful magnetic fields into behaving like gels, or even rubbery solids whose shape responds to changes in the field. This by itself is a kind of crude programmable matter, with mechanical characteristics that can be varied on demand. Such materials are employed in certain types of dampers, shock absorbers, actuators, clutches, and valves. They also have a place in the arts, owing to their eerie, otherworldly dancing under shifting magnetic fields. They've been used experimentally as tactile feedback surfaces for virtual reality, and even as real-time programmable lubricant for ball bearings. Electro-rheological (ER) materials are very similar in application, although their composition is usually based on long, polarized organic molecules. Liquid crystals have interesting ER properties in addition to electro-darkening ones.

The same principles can be applied to solids. For example, a reinforced rubber sheet can have considerable tensile strength but almost no compressive strength; it buckles readily under its own weight. Dope that same sheet with particles of iron, though, and hold an electromagnet over it, and it will happily stand at attention. You could then say that its compressive strength had been improved at your command or, alternatively, that the force of the magnet was able to counteract gravity, placing the sheet in a state of tension. Either way, the rubber goes from being noodle-soft to trampoline-stiff. You could build a self-erecting tent this way, and with other particles in place of iron filings, you could even dispense with the magnetic field and stiffen the tent fabric directly with electrical currents.

Similarly, a cylinder of chalk can bear a significant amount of weight pressing down on it, but crumbles easily when pulled. It has virtually no tensile strength. But with MR or ER dopants, we can place the chalk in a state of permanent squeeze, so that even under some tension it holds together. Again, we could say that its tensile strength is programmable, or that the magnetic and/or electric fields are counteracting the tension. The point is moot; the chalk does not break.

Can quantum dots serve as electro-rheological dopants? Maybe. Probably. I can't think of a reason why not. As for MR dopants, there have already been experimental emulsions using nanometer-sized iron and magnetite particles, like the ones found in bacteria and hard drives. Not surprisingly, in the sort of MR devices described above they work at least as well as, and often much better than, emulsions made from larger particles. And if the dream of programmable paramagnetism and ferromagnetism can be achieved, then we could easily end up with materials more strongly and precisely magneto-rheological than anything previously seen. This is unlikely to result in substances tougher than diamond—you'll eventually run into the failure point of the substrate, and your dopants will be dragged right through it like cannonballs through tissue paper. But we can already see some methods for altering the mechanical properties of a solid in meaningful ways, on demand.

"With arrays of quantum dots," Ashoori notes, "you could make artificial materials with any sort of electronic or magnetic properties that you like. I think there are huge possibilities."

6

The Point-and-Click Promise

> Most of the qualities of an atom derive from its electron cloud. This includes most chemical, material, optical, and electronic properties. Thus the term "artificial atom" is not just hype. Out of this we might get, for example, stronger materials made out of unnaturally stretched electron clouds entwined into fibers, by building a suitable quantum dot substrate to hold the stretched clouds.
>
> —Nick Szabo (December 1, 1993)

THE STRONGEST CRITICISM I've heard for quantum dot-based programmable matter is that it's "unnecessary" because most or all of the effects it can achieve can also be had by other means. Quantum dots have unusual electrical and optical properties, but so do other materials. Quantum dots trap and manipulate individual electrons, but so do large organic molecules such as chlorophyll. Quantum dots mimic atoms, but not nearly as well as real atoms do. Quantum dots can be used to make switches and computers and magnets and heat pumps, but we have all these things anyway. Quantum dots do not, at first glance, bring anything new or "interesting" to the table.

I find this view surprising, especially given the stature, intelligence, and imagination of some of the people who've voiced it. The real power of quantum dots is not that they can be shiny or transparent or insulative, or whatever else we happen to need at any particular moment. The real power is all the *other* things they can be asked to do and be on a microsecond's notice:

Transparent ⇔ Opaque
Reflective ⇔ Absorptive
Electrically Conductive ⇔ Electrically Insulative
Thermally Conductive ⇔ Thermally Insulative
Magnetic ⇔ Nonmagnetic
Flexible ⇔ Rigid
Luminous ⇔ Nonluminous

In theory, a block of truly programmable matter should be able to select any point on any of these axes at any time. Throw in your choice of color and the result is, to my mind, very interesting indeed. And of course there may be other effects such as magnetoresistivity, photo/thermo/piezo-electricity, superconductivity, and the ability to perform classical and quantum computations.

Unfortunately, many of these traits are correlated, so that (for example) a material that managed to be both electrically conductive and thermally insulative would probably face severe restrictions in its other properties. It might be impossible to make it also magnetic, or photoresistive, or green. The choice of substrate would also undoubtedly affect the programmable range of the material. Still, within the limits of physical law it should be possible for a semiconductor like silicon or cadmium selenide to transform itself almost beyond recognition, in a variety of amazing and enormously useful ways. This is a Clarke's Law type of technology, bordering on the magical.

This is where I come in. My involvement with the world of quantum dots began innocently enough, with a smattering of Internet rumors in 1993–94. Something about "quantum dots" and "artificial atoms"—possibly a joke or hoax or naive speculation, since the Internet of the early 90s was full of these. Still, I found the idea intriguing, and mentally filed it away for future reference. A few years later, I encountered these terms again in Richard Turton's authoritative reference, *The Quantum Dot*, and realized it was no joke. It was in fact a Big Deal, and would almost certainly lead to powerful new technologies.

Turton's book itself, though, focused on transistors and had little to say about the other uses of quantum dots, aside from the obvious applications in quantum computing, and certain optoelectronic devices such as infrared detectors and surface-emitting lasers. The text was strangely deficient in speculation of any kind—a fact that turned out to reflect an industrywide bias on the part of quantum dot researchers. The truth is, they were onto something so huge that they were afraid nobody would believe it. Rather than risk their credibility or funding, they were keeping their mouths firmly shut. What Turton did mention, in a quick, offhand way near the end of the book, was this:

> By placing two electrons on a dot we can create a "helium" atom . . . or we could have three or four electrons—in fact any number up to several hundred. Designer atoms can be used to produce an almost inconceivable range of new materials. For example, if several of them are placed in close proximity, the electrons on the dots interact to produce designer molecules.

By profession I'm an engineer, a science fiction writer, and a journalist. All three jobs have in common an obsession with the future, so extrapolating scientific trends into future technologies—while imagining and transcribing the results—is both my job and my main amusement. The future is where I commute to every day. So a statement like this one, buried in the minutiae of a deathly sober text on solid-state physics, really caught my attention.

I started reading up on the subject, everything I could find—which wasn't much. But what I did read quickly led me to the conviction that programmable matter, not computing or optoelectronic gadgetry, was ultimately the killer application. And what an application! I wrote a science fiction story about this, a kind of fairy tale in which programmable matter took the place of traditional magic. By the time it was published in the (now sadly defunct) magazine *Science Fiction Age*, I was already expanding it into a novel. And since the story had a somewhat humorous tone, and many readers had taken this to mean that I was joking about the physics as well, I included an appendix at the back of the novel, explaining a bit about the actual science of quantum dots, and what it implied for the future. And since I'm also a journalist, parts of this appendix were fleshed out for an essay in the British journal *Nature*.

The rest has been like a slow-motion explosion. There were award nominations, leading to more articles and more stories, more attention for the idea, more details worked out. I started getting phone calls and letters and email, some favorable and some very critical of the idea, or of specific aspects of it, of the language I'd used to describe it, or of gaps in my technical insight. In science, this process is known as "peer review," and it's considered not only healthy but vital. Peer review challenges an idea, shines the stage lights on it, pokes it full of holes in the most impersonal and inexorable way. An honest scientist will defend the portions of an idea that work, and abandon the portions that don't. Very few ideas emerge unscathed; some are destroyed, while others mutate and harden into mature scientific concepts.

I'm not a scientist, but I am an engineer—I design and invent, and concoct solutions for weird and sometimes seemingly intractable problems. The peer review process did mutate my ideas about programmable matter, and it also forced them to become more detailed, more defensible, more easily and compactly explained. Somewhere along the way, my business partner, Gary E. Snyder, decided that an article I was about to publish was detailed enough to constitute a patentable invention. At his insistence, we filed an application with the United States Patent and Trademark Office, and within a few weeks we'd been contacted by the U.S. Air Force about the possibility of maybe licensing it.

This was not a career turn I'd expected to take, but discovery is like that. Like falling in love, it comes out of the blue, with its own strange and insistent demands. For anyone interested, the original patent application is included in Appendix B of this book. But what the application says, stripped of legalese, will be the focus of this chapter.

The electrostatic quantum dots employed by researchers like Kastner and Marcus are built on chips that are often capable of generating several artificial atoms at once. This is done for convenience, so that several experiments can take place side by side in the tight confines of a dilution refrigerator. The spacing of the dots is such that the artificial atoms do not interact in measurable ways. In fact, while quantum dot particles are often found in ordered arrays, I'm not aware of any experiment that has placed more than a handful of individually programmable quantum dots in mesoscopic proximity. This is partly because current research is still working out the details of the individual atoms, and also largely because our manufacturing techniques aren't up to the task. But it isn't difficult to imagine, say, a 1-millimeter-square microchip covered with a close-packed grid of electrostatic dots. The properties of Moungi Bawendi's "2D artificial solids" can only be adjusted *en masse*, but this designer chip would be the real McCoy: a slice of matter that is fully programmable at the atomic level.

Now, spacing of the quantum dots on our chip is problematic, since they need to be close enough to interact but not so close that the electrodes of one will have a major disruptive effect on the contents of its neighbors. For argument's sake, let's say the whole thing, electrodes and confinement space and safety margins around the outside, is a square 20 nanometers (~100 silicon atoms) on a side. This size should permit the sort of room-temperature and visible-light interactions we're most interested in. It also means the 1-square-millimeter chip will hold fifty thousand rows of fifty thousand dots each, or 2.5 billion (2.5×10^9) artificial atoms.

Let's further suggest that for maximum flexibility, each quantum dot is of the sort shown in Figure 6.1, with eight electrodes on its top face and eight more on its bottom. This configuration not only allows us to pump electrons in and out of the artificial atom, but also gives us good

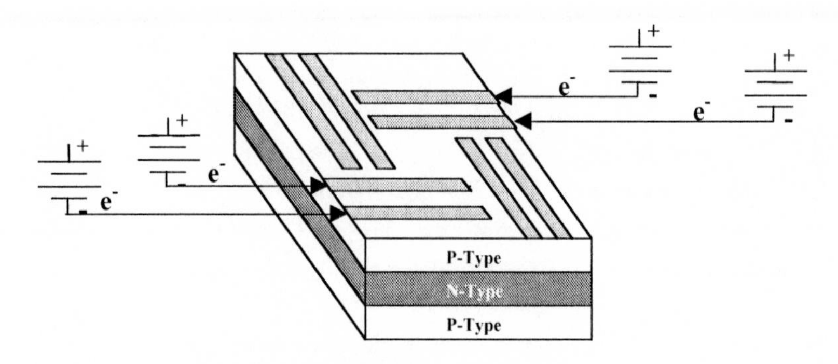

FIGURE 6.1 QUANTUM DOT REVISITED

Pictured here is the notional quantum dot from Chapter 4. Eight electrodes on the quantum well's top face and eight more on the bottom permit precise control over the artificial atom's size, shape, and atomic number. Tiling a microchip with such devices would yield a programmable doping of its surface.

control over its three-dimensional size and shape. What we have, then, are sixteen electrodes per dot, each with an independent voltage source—which means sixteen separate conductor traces feeding into the chip for each of our several billion dots. That's a lot of wires and a lot of voltage sources—40 billion, to be precise. Impractical? Probably. An obvious simplification is to break up the grid into smaller "tiles"—say, groups of 8 × 8 or 64 quantum dots. (See Figure 6.2.) Each of these 64 would be controlled independently of the others, but each tile of 64 would pass the same control signals along to its neighbors, so that every tile on the chip would behave the same as every other tile.

This may sound like a limitation, but if each electrode can be set, for example, to 256 different voltages, then each designer atom will have 256^{16}, or 3.4×10^{38} possible states. Compared to the 92 unexcited states of the periodic table, this is a staggering number, and if we place three designer atoms together, the number of states climbs to 1.02×10^{115}, or 1 quadrillion googol. Since most calculators are incapable of processing numbers above 1 googol (9.99999×10^{99}), this meets my definition of

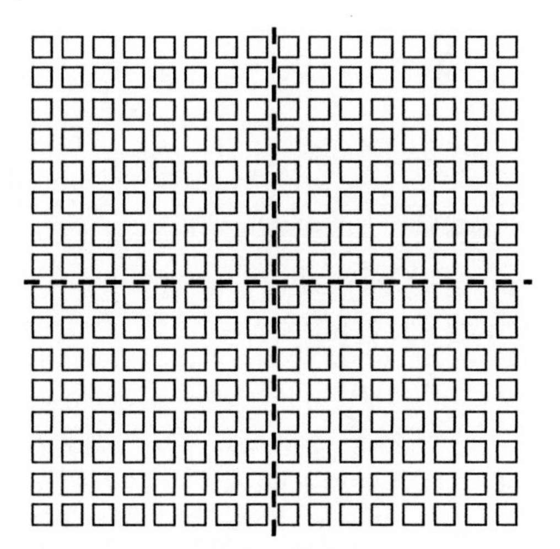

FIGURE 6.2 PROGRAMMABLE TILES

Programmable tiles consist of 8 × 8 groups of quantum dot devices, roughly 160 nanometers on a side. If the dots each have 16 electrodes, then controlling them in groups of 64 requires only 1,024 independent voltage sources. Since multiple tiles can receive the same signals, a surface of arbitrary size can be controlled with only "one K" of input wires.

"effective infinity." If some fantastic materials processing machine were to activate and test a million trillion of these sample states every second, it would take almost a googol years to test them all. So an 8 × 8 grid—more than twenty-one times as large—represents an absurd and downright spooky wealth in undreamed-of materials. Finding needles in that cosmic-scale haystack will be the work of lifetimes.

Controlling the chip itself, however, is relatively easy: with 16 electrodes for each of 64 designer atoms, we have a total of 1,024 or "one K" signals to worry about. If each of these signals is an 8-bit voltage, then we need only 1 kilobyte of memory to represent the commanded state of the chip. Since fitting 1 kilobyte into 1 square millimeter of semiconductor was a trivial exercise even twenty years ago, we can simply add yet another layer to the bottom of the chip, containing 8,192 transistors to serve as bits, and passing 1,024 signal traces up to another layer, which parses them for routing to the individual quantum dots.

Such a chip would be impossible to manufacture using today's technology, and even in the future it will present formidable engineering challenges. The electrical resistance of a wire increases as the wire becomes smaller, so with traces on the order of 1 nanometer wide, parts of the chip could require high voltages (conceivably hundreds or thousands of volts) in order to operate. And with high current, metallic nanowires are subject not only to damaging heat but also to a phenomenon called "electromigration," which can kick atoms out of position. But these strike me as surmountable problems, especially if we can find ways to use diamondoid or fullerene materials in place of metal.

Laying a few hundred of these chips out side-by-side will result in exactly what I promised earlier: a TV screen that changes not only color but substance. With minuscule power consumption, it could easily switch from lead to gold and back again, many times per second. And since it isn't limited to the 92 natural elements, it will be capable of taking on characteristics that natural substances can't. It's a reasonable bet that there'll be better superconductors than today's yttrium barium copper oxides, better reflectors than silver, better photoelectric converters than silicon, and so on, and so on.

Really, such chips would be capable of doing and being so many different things that it's helpful to reiterate the list of their known limitations. They can't change their mass. They can't change their shape, although they can presumably be mounted on the surface of something that can. They can't create or destroy energy, or operate with perfect thermodynamic efficiency. Also, while their "chemical" properties are real, they're weak, and limited to highly localized regions on the chip. At best, you'll have an atomically thin programmable dopant layer sitting near the top of a bed of silicon or gallium arsenide. At worst, you'll have

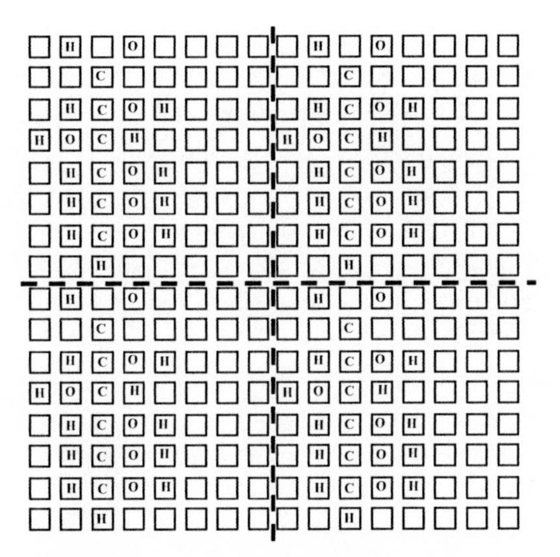

FIGURE 6.3 MAKING IT SWEET?

These phony glucose molecules are too large and chemically weak to trigger a taste bud, but they demonstrate the sort of complex materials we can arrange on a surface of programmable tiles.

discrete programmable islands jutting up from the substrate like stones in a Japanese garden.

We could tile the chip's surface with ersatz glucose molecules, but these would be so oversized and underreactive that our taste buds would never recognize them. (See Figure 6.3.) Still, if we really want the chip to taste sweet, or sour, or like filet mignon, there is enough binding energy available that future engineers may find some dot settings that weakly approximate it.

Wellstone: A Logical Endpoint

A final and very important shortcoming of this technology is its lack of 3D structure. The programmable layer is a nanoscopically thin veneer just below the surface of the chip, and is capable of mimicking only two-dimensional molecules. This rules out the vast majority of organic substances, inorganic crystals, and nanoelectronic devices. You can't command a diamond coating to appear on the chip, or even a quartz one, because these substances rely on three-dimensional structure for their properties. Fortunately, this limitation also has a rather simple solution: we roll the chip around into a long, thin fiber. With the P and N and P layers of the quantum well, and the conducting traces on top of them, and the memory and insulation layers beneath, this fiber would have a thickness of around 60–80 nanometers (300–400 atoms), meaning we could fit 10–13 artificial atoms around its circumference and a potentially infinite number along its length. (See Figure 6.4.)

The artificial atoms are a surface feature of the chip, but by placing them instead on a fiber, we create a material that is *mostly* surface, yet equally controllable. And once we have these fibers, we can embed them in bulk materials to serve as programmable dopants. Perhaps more important, we can string them up in a three-dimensional lattice, not unlike the skeleton of a building, or else weave them together like basket wicker. This is a tough nanoassembly job either way, but once it's complete we have artificial atoms bumping right up against one another, able to bond with neighbors on the same fiber or adjacent fibers. Now we can create not only a thin film of goldlike pseudomatter, but a three-dimensional solid with pseudogold dopant atoms on the inside as well. Thus, we can generate a bulk material with the *mass* of wickered silicon but the physical, chemical, and electrical properties of an otherwise-impossible gold/silicon alloy.

Or we can create mixtures of other metallic or nonmetallic substances, including the "unnatural" ones discussed above. And with the flick of a bit, the voltages on the quantum dots can be altered, to change the solid from one miraculous pseudosubstance to another. In a heated discussion on this subject in the hot, hot summer of 1997, Gary Snyder and I coined the name "quantum wellstone" (or simply

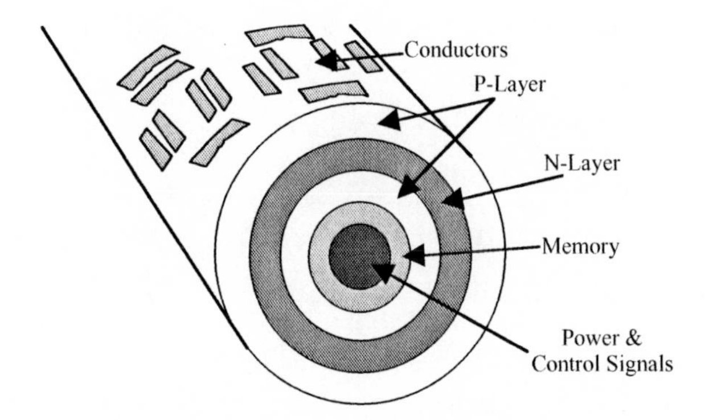

FIGURE 6.4 WELLSTONE FIBER

A notional "wellstone" fiber uses careful arrangements of conductors, semiconductors, and insulators to produce a long, flexible cylinder whose surface is studded with artificial atoms. When woven together, these fibers create a bulk material whose properties are programmable via external signals.

"wellstone") to describe this hypothetical but quite plausible form of programmable matter.

This material's mechanical strength would be based mainly on the substrate material and the size and weave of the fibers, although van der Waals bonding between fibers would also be significant. Tunable covalent bonds are also available, but would be too weak to influence the mechanical properties significantly. In all probability, the material would resemble a polymer such as high-density polyethylene, which also consists of molecular fibers woven (well, jumbled) tightly together. Wellstone would likely be waterproof and airtight, although small molecules such as hydrogen and helium would be able to slip through the weave. It would also be very strong in tension, and bulk samples of it would have a high compressive strength as well. Thin rods or sheets of wellstone would be extremely flexible, although magneto- and electro-rheological forces would be available to stiffen them if necessary.

The Real World Intrudes

Alas, simply defining the characteristics of wellstone does not make it real. There remains the daunting question: how do we manufacture it? Thanks to a physical barrier called the "limit of diffraction," photolithography can't easily produce structures smaller than a typical wavelength of ultraviolet light, about 250 nanometers. Think of this as the "pixel size" for the imaging process that etches the traces, although structures as small as 100 nanometers have been produced with heroic (and expensive) efforts. Light waves with higher energy and shorter wavelengths than ultraviolet would be helpful, but unfortunately this puts us into the extreme UV and x-ray portion of the spectrum, where normal materials can no longer reflect and refract the beam in the ways we would like. X-rays are also dangerous; both people and materials suffer health problems with too much exposure. Still, advancing technology in these areas might eventually enable the printing of features 10 nanometers across, or even smaller.

Electron beam lithography is more promising, in that it can etch lines only a few nanometers thick, and maybe someday all the way down to atomic scale. The beam must etch its lines one at a time, though, rather than all at once as with photolithography. Therefore, electron beams are generally too slow for mass production and are useful mainly for making research prototypes. A number of techniques are under development for atom-by-atom assembly of nanostructures, but they suffer from much the same problem: any assembly operation that is not massively parallel is going to be really, really slow by the standards of today's electronics industry.

Chemistry provides another possible avenue. Aided by the ability of some structures and materials to self-assemble at the nanoscale, chemists have proven increasingly adept at producing and manipulating complex designer molecules. Chemists growing "striped nanowire" structures of metal and semiconductor have already produced working diodes, LEDs, transistors, and crude circuits as small as 1.5 nanometers across. Others have used designer viruses to assemble three-dimensional structures of colloidal quantum dots. Most amazingly, in mid–2002 a team at Harvard and Cornell managed to use the electromigration of atoms to create transistors out of single atoms. The wires running into the device are much larger, of course, but Cornell's Paul L. McEuen notes that these atom-scale transistors represent the ultimate limit for quantum dot technology: electrically shrinking or growing the swarm of electrons surrounding a single real atom. So while practical nanocircuitry is beyond the reach of today's chemists, their bag of tricks is growing rapidly.

I have confidence in the ingenuity of human beings, especially when there are fortunes to be made. If these nanostructures are as useful as expected, I think it's safe to assume that some means will be found not only to produce them but to *mass*-produce them for the consumer market. Perhaps we'll see hundreds of nanoassembler arms and electron beam generators packed onto a chip, which can build up layers of wellstone as though they were pizzas, and stack them endlessly, one upon the other, to extrude a fiber of arbitrary length. These nano-silkworms might even have the power to crawl along pre-programmed pathways, in a nanoscale spinning and weaving operation that turns the fibers directly into something like wellstone cloth, or large blocks of bulk material.

Unfortunately, wellstone faces other problems as well. Atoms and molecules from the air, powered by the thermal excitation we call "temperature," constantly bombard any surface. Atoms can be knocked loose from an exposed nanostructure, and foreign atoms deposited where they do not belong. Exposure to air, water, and other contaminants also affects the electrical and mechanical properties of nanoscale materials, so that they do not behave in ideal or cleanly predictable ways. This is why atomically fine structures in the laboratory are kept in vacuum rather than air, or else stored and operated at low temperature to prevent damage.

To be useful in the real world, though, programmable matter has to operate in normal air, at room temperature and above. This consideration will certainly affect wellstone's design and composition. For example, it probably places a lower limit on the size of the fiber, so that its exposed structures are large and coarse enough to suffer the indignities of atmosphere. Alternatively, the entire fiber might be encased in a protective sleeve. Fortunately, the 3D crystal structure of semiconductors makes them pretty tough—it's the conductive electrodes we really have to worry about. Still, if these are made of carbon nanotube or something similarly durable, then we won't have as much to worry about.

A final problem for nanoscale structures is that their ratio of surface area to volume is extremely high. Thus their behavior is dominated by the surface properties, rather than the bulk properties, of the materials they're made of. This feature is critical to the functioning of wellstone, but it also means that the normal rules for electronic devices do not apply. We're out of the microscale here and into the murky lower depths of the mesoscale. For example, the control wires embedded in a wellstone fiber could conceivably be only a few atoms wide, meaning that virtually every atom in the wire would be either at or adjacent to the wire's surface. If the wire were a hollow carbon nanotube, then *all* of its atoms would be surface atoms. Our current science does a poor job of predicting the behavior of such wires, so there may be hidden problems here as well.

As physicist and author Robert A. Metzger puts it, "The top 3–5 nanometers of a clean semiconductor wafer is mostly oxide, full of junky unsatisfied bonds. It can store a tremendous undesired charge." This of course has quantum effects of its own, and needs to be taken into account when designing the outer sleeve of a wellstone fiber. Diamond may be a highly desirable material here, as its surfaces acquire an atomically thin coating of hydrogen from the atmosphere, rather than oxidizing.

Other problems may be alleviated if we replace the central control wires of a wellstone fiber with optical conductors, or perhaps with electromagnetic transceivers that, due to their size, would likely operate at extremely high frequencies. Any method for reliably transporting data and energy is a potential candidate for use here. We don't have to know all the answers right now, but it does help to get our arms around the questions.

Surface properties raise another interesting point as well. We've previously noted that artificial atoms are at their most atom-like when small, and in general this is a good and desirable thing. However, if the free electrons in a wellstone fiber were confined to regions only 1 nanometer across, they'd occupy only about 0.3 percent of the fiber's surface, whereas if they were 10 nm across they would occupy fully a third of it. So while smaller quantum dots may have sharper spectral lines and more energetic pseudochemistry, *larger* dots may have a greater ability to link up into pseudomolecules that dominate and swamp out the natural behavior of the semiconductor. Today we can tinker with the density, positioning, and atomic number of dopant atoms, but we are powerless to change their size, or even understand what effects this might have. This fascinating tradeoff could well be a hotbed for future materials-science research.

Does wellstone work? There will always be naysayers for any new idea, particularly one as speculative as this. But quantum dots can and do affect the optoelectrical properties of matter, and few researchers will deny there are materials-science implications. And in a three-dimensional universe, rolling the dots into fibers (or connecting them to wires) is *the* way to control them inside a bulk material. Everything else is detail; the underlying principles can't fail to work. But how well is an open question; for a woven quantum dot solid to behave reliably in the real world, there are practical limits to how thin the fibers can be, how much current they can carry, and how many electrons can be crammed onto their surfaces. For a variety of reasons, I suspect the size limit is around 60 nm, although Charlie Marcus has suggested that a fiber as thin as 20 nm might be coaxed into performing some useful tricks.

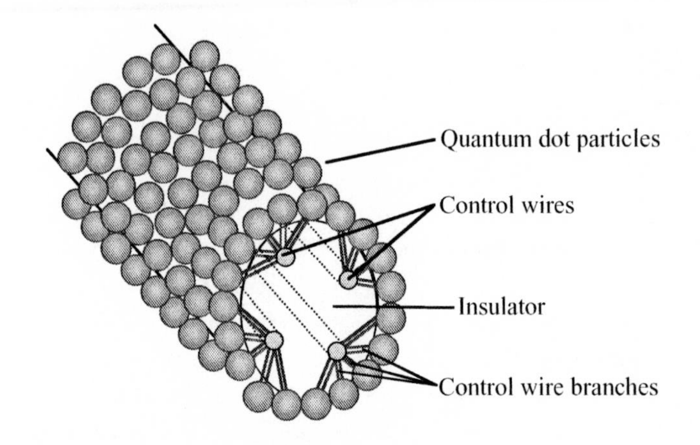

FIGURE 6.5 GRAPEVINE WELLSTONE

"Grapevine wellstone" consists of colloidal dots attached to the surface of an insulating fiber, with control wires running through it and branching out to the surface.

Shortcuts to Wellstone

The daunting smallness of these scales makes it both impossible to construct a wellstone fiber and intractable to model one using today's technology. But there may be some shortcuts that permit wellstone-like behavior from simpler and easier materials. The most obvious of these is a design I call "grapevine wellstone," which replaces the electrostatic quantum dots on the fiber's surface with Bawendi-type particles. (See Figure 6.5.) We already know how to make these, and understand a great deal about their behavior, so this design removes at least one difficult step from wellstone's manufacture. But it also introduces the fragility of a colloidal film, and eliminates any ability to adjust the size or shape of the artificial atoms.

Another, even simpler version—birdshot wellstone—places a colloidal film of quantum dots on the exterior of a single control wire. (See

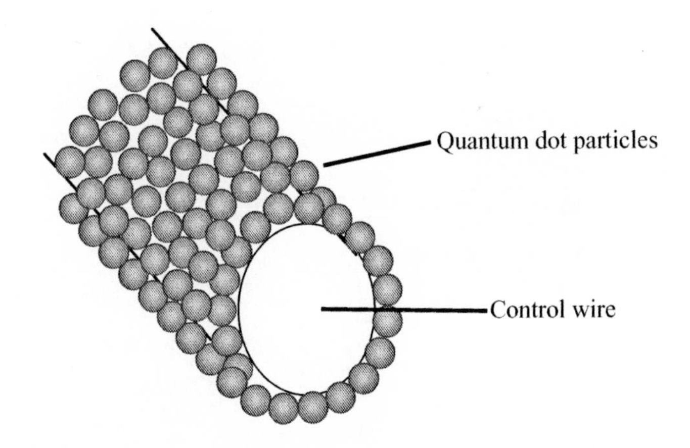

FIGURE 6.6 BIRDSHOT WELLSTONE

"Birdshot wellstone" is possibly the simplest form of programmable bulk matter. It consists of a film of colloidal dots attached to the surface of a metal wire. Touching the wire to ground—or to another wire—forces electrons into the quantum dots.

Figure 6.6.) If a high impedance were placed in series with the wire, or if it were grounded lengthwise to a drain wire, electrons could be forced into the quantum dots as the voltage on the wire increased. There are no important limits on the size of the wire, so it could be fairly large. This arrangement can therefore be manufactured and tested with today's technology. Unfortunately, it limits the fiber's programmability even more severely; as with Bawendi's 2D solids, all the dots on the fiber will have approximately the same number of electrons inside them. Still, it provides a near-term means to get programmable dopants inside a bulk material, and even to form small samples of a woven solid. This would permit an experimental assessment of wellstone's limitations and operating principles—extremely helpful before proceeding to the more difficult steps.

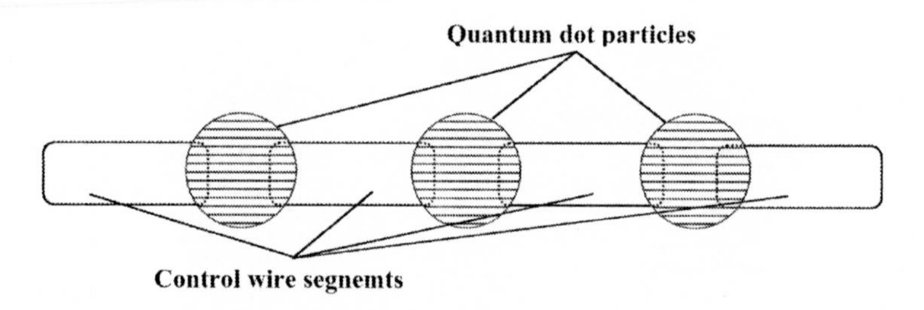

FIGURE 6.7 STRING-OF-PEARLS WELLSTONE

"String-of-pearls wellstone" has a number of drawbacks but can be made arbitrarily small.

Still other options exist. One could, for example, embed control wire segments inside quantum dot particles, forming a string of nanoscopic beads—"string-of-pearls wellstone." (See Figure 6.7.) This version is probably harder to produce than the "birdshot wellstone" described above, and it has the additional drawback of resistive voltage losses inside the quantum dots. The voltage would drop along each successive link in the chain, so that the fiber as a whole, rather than containing many repetitions of a single pseudoelement, would entail a decay sequence in which the quantum dots at the head of the chain contained more electrons than the ones at the far end. Whether this is useful depends on what effect you're trying to achieve, but it will certainly tweak the properties of whatever you embed it in. Other permutations include replacing the control wires with fiber-optic conduits, although this may create additional problems as the conduits become many times smaller than a wavelength of light.

The smallest conceivable wellstone fiber is simply a string of metal atoms with "pearls"—larger atoms or clusters of atoms—spaced every so often. Unfortunately, this design has no structural integrity and would fly apart outside a helium bath. The smallest *practical* fiber would be similar in design to a molecular single-electron transistor (SET) described by Harvard physicist David Goldhaber-Gordon in 1997. This consists

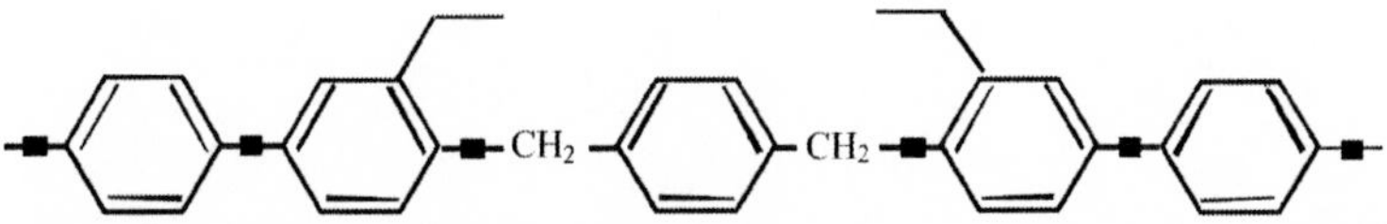

FIGURE 6.8 SINGLE-ELECTRON TRANSISTOR

Goldhaber-Gordon's molecular single-electron transistor (SET) is possibly the smallest quantum dot fiber that will hold together in free space. The quantum dot is the hexagonal benzene ring in the center, surrounded by insulating CH_2 molecules.

of a "wire" made of conductive C_6 (benzene) molecules, with an inline "resonant tunneling device," which is simply a conductive benzene molecule surrounded by CH_2 molecules that serve as insulators. (See Figure 6.8.) This forms a two-dimensional, hexagon-shaped quantum dot that is one-tenth the size of a C_{60} buckyball, and consists of only six carbon atoms. It is intended as a transistor (i.e., an electrically controlled switch) rather than as a confinement mechanism for electrons. However, like McEuen's atom-scale transistor (but with much thinner and tougher wires), the device should be capable of containing a small number of excess electrons and, thus, of forming a primitive sort of artificial atom.

The device would suffer the same limitations as string-of-pearls wellstone, and the small number of confined electrons would limit its usefulness still further, but it has the advantage of being chemically producible with today's (even yesterday's) technology. Bulk crystals of it ought to have some rather strange properties. A somewhat larger design, capable of holding many more electrons, might employ carbon nanotubes as the control wire segments and C_{60} or larger fullerene molecules as the quantum dot particles.

The Matter in General

Programmable matter faces few theoretical barriers, if any. The underlying principles are sound and have been laboratory-demonstrated in limited ways. Furthermore, quantum dots and quantum wells have a solid track record of producing material properties that don't occur in nature. But the practical challenges are huge, and it would be very premature at this point to sign off on the wellstone devices described above. More understanding, both theoretical and experimental, is needed before any design like this can be taken at face value.

Nevertheless, areas such as band gap engineering, single-electron manipulation, designer molecules, molecular nanotechnology, nanoelectronics, and quantum computing are converging on this capability from a variety of directions. Gary Snyder points out that quantum confinement should also allow us to construct atoms out of "holes" as well as electrons, forming a bizarre kind of negative matter. Not antimatter, but something entirely new.

As Sun Microsystems' Howard Davidson puts it, "Quantum dots fall into a very small subset of what might be called programmable matter. Any system that can confine and manipulate an electron can change its properties as a result. With designer materials using all sorts of technologies, not just quantum dots, we'll get all sorts of materials that don't occur in the universe. And with very high surface-to-volume ratios you get these weird optical and chemical properties. There is still an awful lot of room in materials science to produce some very unexpected things."

Indeed. I submit that matter with real-time adjustable properties—in whatever form and by whatever means—should be a primary goal of materials-science research. The question of *why* is best addressed through examples, so for the remainder of this book I'll use the terms "wellstone" and "programmable matter" interchangeably, with the understanding that the actual technologies involved may be somewhat different from the ones described above.

7

The Programmable City

"Full transparency on the roof and east wall, please." Obligingly, a third of the house turned to glass. To actual glass, yes, and if danger threatened it could just as easily turn to impervium or bunkerlite, or some other durable superreflector.

—Wil McCarthy, "Once Upon a Matter Crushed" (1999)

FOR MUCH OF HUMAN PREHISTORY, nomadic tribes lived in leather tents, or in temporary huts woven from local vegetation. Early farming villages were made of wood, and the first permanent cities were built of mud bricks. With the advent of copper tools, it became possible to shape large blocks of stone, which became the preferred building material, and permitted structures substantially larger and more permanent than anything before them. The 4,500-year-old Pyramids are the most striking example, but ordinary houses could be built of stone as well, and often were. The Egyptians also invented mortar and experimented with primitive concrete.

Later, the Greeks introduced the idea of joining stone building blocks not only with mortar but with bronze dowels and butterfly-shaped clamps, which permitted larger structures to be nearly as sturdy as smaller ones made from fewer pieces. The Romans, for their part, invented horsehair-reinforced ash-lime concrete, a material much easier to shape, and yet so durable that many of the bridges and aqueducts built from it are still in daily use some two thousand years later. Concrete isn't as strong as most types of stone, but its moldability and adhesive qualities meant that large structural members could be cast in place as single pieces. This enabled the Romans to build very large, complex structures such as the Coliseum and, more practically, multi-storied apartment buildings to house their millions of citizens. Roman architecture tended to be airy and open, which was sensible considering the local climate.

The Middle Ages saw one significant change in materials, at least for the wealthy: glass windows. This permitted societies in the cooler climates of northern Europe to let sunlight indoors without also letting in cold air. In the energy-rich world of the Industrial Revolution, it became practical for the first time to make buildings with metal skeletons. These could be much taller than brick, concrete, or stone buildings alone; cities began to sprawl upward as well as outward. And when ordinary concrete was replaced with the steel-reinforced variety, it became possible to create buildings so tall and heavy that the ground underneath them would fail long before the buildings themselves. Owing to the lower energy costs of their manufacture and transportation, these materials are cheaper than stone as well as stronger and more versatile.

Soon, with the advent of cheap metal piping, water and natural gas could be brought into buildings, and sewage brought out, sanitarily, beneath the floorboards and streets. And as copper transformed from a precious metal into an industrial one, cheap electrical wiring became possible as well. Our modern cities of steel and neon, fountains and mirrored glass, are thus a mosaic of the inventions of ages past. The lesson: even incremental advances in materials science create huge changes in the way people live. Urban architecture has always been the art of the just-barely-possible, and there's no reason to think this will ever change.

As the twenty-first century unfolds, new materials are becoming available at an ever-increasing rate: foamed metals, ultralight ceramics, plastic wood substitutes. . . . Laboratories are even developing the skills to mass-produce the strongest and hardest of substances: silicon carbide and diamond. It becomes more and more difficult to know what's still impossible. And with programmable matter on the horizon, we can expect further marvels that alter our cities just as radically.

Against the Wind

The properties of wellstone will not be infinitely programmable; whatever settings we feed into the artificial atoms, they will still be surrounded by small amounts of metal and fairly large amounts of semiconductor, which will certainly affect its properties. As I've noted, you can't turn the stuff harder than diamond (although diamond is a semiconductor, and making wellstone fibers out of it is a real possibility). But as long as there is power available, you can almost certainly use electrorheological (ER) or magneto-rheological (MR) properties to stiffen it considerably. In fact, chemists at Stanford University are already investigating the uses of MR materials in architecture—not as walls but as suspension, shock-absorbing, and vibration-dampening elements. Buildings made from stiff materials tend to be brittle, and thus easily damaged in earthquakes, explosions, or heavy wind loads, while overly flexible structures can become unstable and even buckle. The "smart buildings" of the future are expected to tune their stiffness in response to external stimuli. To bend in the wind, yes, but not to flutter or crack.

The major objection to this is obvious: MR and ER materials are stiff only when power is applied to them. Otherwise, they're limp and compressively weak. Who would trust a building that relied on electricity for its structural integrity? The answer is: people who stand to save money. Inflatable tennis domes are one present-day example, and in extreme climates there are buildings that sit on permafrost and actively refrigerate their foundations to keep them from melting. In either case, the building may easily collapse if certain systems break down while the inhabitants aren't looking. Obviously, most buildings aren't made this way, and won't be in the future either. Still, wellstone may be an important part of future architecture, thanks to its many other useful properties. Its ability to control the structural dynamics is a freebie. Buildings would probably need some minimal skeletal support to survive power outages and such, but again thanks to programmable matter, such outages should be rare indeed.

Follow the Sun

The power of full equatorial sunlight, at noontime on a horizontal target, is around 1,100 watts (1.1 kilowatts) per square meter, which is really a hell of a lot of energy—more than enough to cook a pot roast or to kill an unprotected person in a couple of hours. In a more practical vein, and accounting for minor details such as nightfall, weather, seasonal variation, and latitude, the sun deposits an average of 1.5 to 3.0 kilowatt-hours per square meter (kWh/m^2) every day in temperate climates. The kWh is a standard measure of energy, and households in the developed world today consume, on average, about 80 of them per day.

So in principle, a roof made of 100 percent efficient photoelectric materials would need an area of only 53 square meters (less than 22 by 22 feet) to meet the energy needs of a typical family, as opposed to 442 square meters for a standard commercial (amorphous silicon) solar voltaic system today. In practice, though, the efficiency is affected by foliage, dust, the angle and geometry of the roof, and the storage medium for the energy.

Since homeowners prefer to have power on demand, rather than when the sun happens to provide it, energy is typically stored in batteries. Programmable matter can probably help us here as well. In an ideal universe, it would provide us with room-temperature superconductors capable of carrying enormous currents in very tiny volumes. Failing that, it should at least be capable of serving as a capacitor. If our hypothetical house were made mostly of wellstone or some close cousin thereof, its walls could serve as both wiring and energy storage—scaleable on the fly, to meet the household's changing needs as the hours and days and seasons unfold. In times of high consumption, virtual

TABLE 7.1 DAILY HOUSEHOLD ENERGY CONSUMPTION (USA, 1993–1997)

Space Heating	41 kWh
Water Heating	15 kWh
Refrigeration	7 kWh
Space Cooling (air conditioner)	3 kWh
Lighting	3 kWh
Clothes Drying	2 kWh
Cooking	1 kWh
Dishwashing	1 kWh
Other Appliances (TV, stereo, computer, etc.)	7 kWh

plaster would give way to fat conduits of pseudocopper, and to capacitors of pseudocarbon and glass.

If we pessimistically assume a 40 percent overall efficiency for the system, then a roof area of 134 m^2 (roughly 35 by 35 feet) would be sufficient to meet most homes' energy needs even in a particularly cloudy year. In the gloom and chill of Fairbanks, Alaska, where the annualized insolation can be as little as 0.2 kWh/m^2 per day, it's much harder to run the entire house on sunlight. Still, this same roof would shave more than 13 percent off the annual utility bill. And keep in mind, 134 m^2 is the size of a mobile home; the average house, with average energy consumption, is probably closer to 160 m^2, and areas in excess of 250 m^2 are by no means unusual. So bigger homes would provide not only more living space but more energy as well. Large homes in moderate, sunny climates could easily run a surplus, and even sell energy back to the power grid, so that in the future, as in the present, the rich can get richer.

The poor, however, can get richer too, as we can see with a closer look at how those daily 80 kWh are used (see Table 7.1). Even a quick glance at this table reveals one critical fact: most of the energy in a home is spent heating things up and cooling them off. A typical house is doing both things at once: chilling food in the refrigerator at 35 percent efficiency, which dumps waste heat out into the living spaces, which are then cooled off with an air conditioner, while somewhere downstairs another heater is keeping 40 gallons of water at just a few degrees below boiling, even on the hottest days of summer, even when no one is home. When it comes to thermal management, the modern house is a mass of contradictions. Wouldn't it be nice if we could simply move heat around to where it was needed, and away from the places it wasn't?

That Warm Glow

Heat travels from one area to another in several different ways. The simplest is by radiation, in which heat travels as invisible, long-wavelength infrared (IR) light waves. These waves can be absorbed and reflected just as visible light can. A black surface exposed to them will absorb heat, and grow warmer; a mirrored surface will absorb much less, so its temperature will hold steady. A shiny surface also reflects an object's own heat back into its interior. Conversely, when they're placed in shade, dark surfaces can also serve as efficient radiators, dumping heat overboard as infrared. This principle works best in high-altitude, low-humidity areas—it's astonishingly effective in outer space, where it can rapidly generate temperatures cold enough to liquefy most gases. But even in the dense, humid climates of sea level, where 75 percent of the human race lives, it will still have an effect. And remember, even where the effect is small, every watt-hour the house can radiate away will save at least one watt-hour in active cooling.

Thus a house that's black in the sun and silver in the shade gets very hot in the summer sunshine, while a white or silver one with deep black shadows remains cool. But in the winter, sunshine keeps that same black house nice and warm, while the silver one is considerably less so. One obvious application for programmable materials is therefore to change color in response to temperature (thermochromicity) or, better yet, to change color on demand, in response to a variety of inputs including weather, temperature, neighborhood covenants, and the tastes of the owner. Of course, an LCD screen can already do this, and many people have asked me whether I believe this makes the LCD a form of programmable matter. The answer is yes, absolutely, although it's a very primitive sort, with large electronic components and only a handful of adjustable properties. Anyway, a house whose exterior was one giant LCD could not only change its appearance but regulate its temperature by at least ±10°C. You could also show movies on it, or even turn the house "invisible" by showing real-time video of the view on the other side of the building. How cool is *that*?

Unfortunately, it takes a fair amount of power to run an LCD, and even more to run a backlight. This is one reason cellphone and laptop displays are switching over to arrays of light-emitting diodes. These are brighter, more efficient, and more durable, and should eventually be cheaper as well. Unfortunately, they're purely emissive devices that are of no use in changing the reflectivity and absorptivity of materials. Quantum dots, on the other hand, have already shown great promise in these areas. As we've seen, they can also change *transmissivity*, meaning that surfaces can be made transparent as well as opaque or reflective.

This trait could be very useful in the design of windows; today we make them out of silica-soda-lime glass, which is transparent to visible light but reflective to the longer wavelengths of infrared. This creates the familiar "greenhouse effect," in which the materials inside your house or car (or greenhouse) absorb visible light, heat up a little, and release the excess energy as infrared light, which then can't escape. This effect is wonderful in the wintertime, at least during the day, when glass houses may not require any heating at all, but it's terrible in the summer. A tall skyscraper can easily absorb a megawatt (1 million watts) of solar heat, which then has to be pumped out with inefficient, power-hungry air conditioning systems. With quantum dots, it will likely be possible to create "glass" that is as transparent to infrared as it is to visible light. Even more significant would be a material whose transparency was tunable at multiple wavelengths, allowing precise control over the light and heat that enter and leave through the windows.

Taming the Flow of Heat

In cutting domestic energy use with programmable matter, the refrigerator—the number 3 consumer of household energy—is an easy place to start. On a cold day, it makes little sense to expend energy keeping food cold when we can simply use the outdoors as a heat sink. Imagine that the refrigerator is mounted into an exterior wall of the house, and that instead of a freon compressor and radiator on the back, it simply has a big block of metallic silver, which extends through the wall and outside, to a flat, refrigerator-sized silver panel.

Silver is an excellent thermal conductor: a square meter of it, ten centimeters thick, will conduct just over 4,200 watts of heat for every degree of temperature difference across it. If the inside of the refrigerator is at 3°C (38°F or 276°K), and the outside temperature is 0°C (32°F or 273°K), then the block will carry heat from the inside to the outside at the staggering rate of 12,600 watts. To get an idea of just how much heat flow this is, try grabbing a 40-watt light bulb with your bare hand. Now multiply that discomfort by a factor of 315. That's a lot of energy being transported, and will very rapidly equalize the refrigerator temperature and the outside temperature, at zero energy cost.

There are also carbon-fiber materials that can conduct 25 percent more heat than silver, and diamond is nearly *five times* better. But if we assume that the best wellstone can achieve is the thermal conductivity of undoped silicon, we will still be channeling 3,750 watts of continuous heat. More than enough, by far. Unfortunately, on a hot day this same thermal shunt will channel heat *into* the refrigerator, which is precisely why we don't build refrigerators this way. But on those days, if the wellstone block could change its electronic structure to resemble an insulating substance such as quartz (SiO_2), we could cut the flow to only 1 percent of that rate, keeping the heat outside where it belongs, at least for a while.

Of course, even if the refrigerator's insulation were perfect (thermal conductivity of 0.000000 w/m-K), heat would still leak in every time the door was opened. So really, it isn't a refrigerator at all, but an icebox of the sort found in American homes until the 1950s, where cooling is accomplished by dumping in things that are already cold. In everyday use, an icebox requires a steady supply of ice, which is a nuisance and requires regular trips to a depot or regular visits by an ice-delivery service. If this were desirable we would all still be doing it. What we really want and need is a real refrigerator, which takes in power from some source and uses it to pump heat away and keep its interior cold. Fortunately, with programmable matter we can harness the Peltier effect on demand, keeping that same thermal shunt in place as a heat sink, but inserting a thermoelectric layer between it and the interior of the fridge. Cooling is inherently an inefficient process, but with the improved materials already expected, the energy consumption under even the worst circumstances should be comparable to today's mechanical refrigerators, while the average, annualized consumption is a tiny fraction of that.

These same principles apply to the cooling of the house's interior: through a switchable combination of photovoltaics, conductors, insulators, and Peltier junctions, we can accomplish this using nothing more than the walls themselves. In fact, for thermal management of an entire house, wellstone offers a number of additional options that may, under some circumstances, cut the energy bill even further.

The easiest and most obvious of these is to dump excess heat down into the ground when the house is too hot, and draw it up out of the ground when the house is too cool. Thanks to escaping heat from the Earth's molten interior, the ground a few meters below the planet's surface maintains a steady, year-round temperature of around 55°F (13°C), regardless of weather. It gets much hotter a few kilometers down, but that's not much use to us architecturally. Still, a building with heavily insulated walls and ceiling and a thermally conductive foundation extending three or four meters into the ground will maintain that same 55-degree temperature inside the home.

That's a little cool by human standards, alas, and the ground itself is also a poor conductor of heat, meaning the area just beneath the house may absorb heat or cold faster than it can transport it to the surrounding dirt or rock. Over time, especially in extremes of hot or cold, we'll saturate its ability to regulate the house's temperature. Anyway, a house can't be completely sealed against the outside without suffocating its inhabitants. Some sort of ventilation is necessary, bringing outside air in and letting inside air out. Also, a human body generates, on average, about 100 watts of heat, with peaks of up to 500 watts while performing hard work. Various household tasks such as cooking and laundry will inevitably generate additional waste heat, so even if your floors can switch on demand from quartz to silver and back again, you'll still have some temperature problems to deal with.

Still, with even a modest combination of thermochromicity, IR transparency, and ground-source heating/cooling, we can cut the overall energy bill enormously.

Heat and Lighting

Lighting a house also presents a number of thermal problems: windows trap heat, and electric lighting generates it. In the winter this is no problem—it's even desirable—but in summertime it simply adds to the heat burden of the house. With programmable matter, the smart house of the future will be able to create a window wherever it's needed. Photodarkening and electrodarkening materials have been around for decades, and quantum dots have already demonstrated the same effects. In fact, with entire walls and roofs and ceilings of such materials, the very concept of "window" becomes outmoded; houses will simply turn transparent in certain areas, in ever-changing patterns depending on the thermal and luminary needs of the inhabitants. Instead of flipping on electric lights, we may someday flip on windows and skylights, and of course adjust their visible and IR transparency while we're at it.

This doesn't light things up for us in gloomy weather, of course, or at night. Unfortunately, incandescent light bulbs rely on a tungsten filament heated to over 2,500°C. Although it's white hot, it releases more than 75 percent of its energy as invisible infrared. A 40-watt bulb creates at least 30 watts of heating directly, and much of the remaining visible light is absorbed by surrounding objects, heating them further. A candle flame is even worse: it hovers around 1,700°C, which is only yellow-hot, and over 90 percent of its emissions are infrared. These are extremely inefficient light sources.

Fluorescent lights, video phosphors, and light-emitting diodes operate on a completely different principle: rather than releasing light as the high-frequency end of a thermal glow, they simply excite electrons in room-temperature atoms. Depending on your point of view, this either elevates the electron to a higher energy, and thus a higher point above the nucleus, or else it creates a nearby "hole" particle that can flow and propagate through the material much like an electron. When the electron relaxes back into its natural energy band—or recombines with a hole—it releases that excess energy in the form of a photon. The emission spectrum of a hot object is continuous—it releases photons at all wavelengths, centered around a peak that increases with temperature.

The spectrum of a fluorescent object is discrete—it contains narrow peaks centered on the wavelength of specific electron energies. These sources produce far less heat and consume far less energy than incandescent lights of similar output. They also tend to produce light in specific colors. Today's "white LEDs" are actually blue or violet LEDs shining through a layer of fluorescent white phosphorus. Combined with a slightly higher cost (though much longer lifetime), this lack of a broad output spectrum has prevented the widespread acceptance of LEDs in our homes.

Once again, though, programmable quantum dot materials offer a solution. An individual quantum dot can produce light only at one specific frequency (or color), which is determined by the energy levels of its trapped electrons. But an array of quantum dots (say, 8 by 8) could produce dozens of different colors, their narrow individual frequency bands smearing together to match the color spectrum of any desired light source, including sunlight, incandescent or fluorescent lights, candle flames, and so on.

This light would emanate directly from the walls and ceiling of the house, either in discrete, movable, highly directional sources resembling spotlights (or even lasers), or as a more diffuse glow. The floor, for example, might cast photons upward in attractive (and movable) patterns. A countertop or workbench might serve as its own illumination—impractical but interesting. Or perhaps the whole interior of the house would emit light, providing a low, sourceless glow, making silhouettes of the furniture and people inside. The aesthetic possibilities are virtually limitless.

There is no reason why this mix of light shouldn't also include thermal infrared. Radiant heating is an efficient and popular means of staying warm, and when the heat source can be moved anywhere (including everywhere) on a moment's notice, it may be the best possible method for avoiding the hot and cold spots that cause so much annoyance and strife in today's homes and offices. And unlike the production of light, the process of converting stored energy to heat is essentially 100 percent efficient, because any losses or inefficiencies in the system bleed energy off as waste heat anyway.

So, rich or poor, big or small, the programmable matter home is a marvel of energy efficiency, consuming far less than 80 kWh per day. In sunny climates it's fully self-sufficient, and even in cold or gloomy ones it should require minimal energy from nonsolar sources. A simple 100-gallon propane tank might last for months, or even years.

Better Homes

Imagine a little house in the desert, perhaps in Arizona or eastern California, where a man and a woman—whom we'll creatively call Mom and Pop—live with their young son, Sonny. It's 5:59 A.M., just after sunrise, and the house is one big, dead-black solar collector. Its exterior drinks in sunlight, taking a little of that energy in as heat and pumping the rest into electrical wires, which shuffle it off into a growing bank of capacitors. There are no windows; the house would be completely dark inside if there weren't dim little nightlights glowing here and there, lighting the way in case someone has to go to the bathroom or stumble to the kitchen for a nighttime snack. Beneath the house-shaped solar array, the walls and roof are one big, opaque insulator, keeping the cold, bright morning at bay.

Then, suddenly, the clock ticks over to 6:00 A.M., and an arrangement of windows and mirrors fades gradually into existence, waking each of the inhabitants with the gentle but undeniable kiss of morning sunlight. These windows and mirrors are in a different position every day; over the year they trace out analemmas, or astronomical figure-eights, on the walls and ceiling. So Mom and Pop and Sonny awake, wish each other good morning, then crawl blearily out of bed to bathe and dress for the coming day.

More windows open; the household comes alive. The family eats breakfast in a little solarium, offering them a nearly panoramic view of the sunrise over the low hills surrounding their home. Their privacy isn't compromised; through the mirrored glass, an observer outside would have a hard time seeing anything more than his own reflection. Soon the front door opens, allowing Dad and Sonny to step outside and pile into one of the family's two automobiles. Today is a school day and a work day. Mom, an architect, remains at home and settles into her office for a morning of leisurely but profitable work. Her desk is also her blotter, computer screen, and drafting table. She "writes" on it with any pointed instrument—even her finger will do—and spilling coffee doesn't hurt it a bit.

Mom becomes restless, though, and mulling over some difficult idea, she gets up from her chair and begins to pace through the house. Windows and skylights follow her around, but now they're looking out mainly from the house's shady side, avoiding direct sunlight. The house, knowing that the day will be a hot one, is becoming stingy about accepting more heat. Its energy stores, though, are swelling rapidly, and when

11 A.M. arrives and Mom departs for a business lunch, the house has socked away about as many kilowatt-hours as it can. Storing any more would mean converting visible portions of wall or floor into additional capacitor space—a practice that its programming finds distasteful. The air outside is hot when Mom opens and closes the door.

While she's gone, the house closes all its windows and converts almost its entire exterior into a mirror, keeping only a small solar collector for maintenance voltage. The domestic hypercomputer—with dozens of EM in disposable computing power—idly browses its library and encyclopedia and the Internet, finding news, entertainments, and educational material that it knows its family will enjoy. With the power of quantum computing, it's even able to analyze this information, and to ponder deep questions like the meaning of life and the best routes to happiness. It enjoys giving advice when asked, although it rarely volunteers.

Inside, the rooms are almost completely dark; the cat is napping, and there is no one here who needs light. But the house does need to keep the air fresh, and this lets in heat despite every precaution. At 11:35, the first small Peltier junction appears, and by the time Mom returns at 2:08, the junction and its black, shadowed radiator have grown to the size of a Volkswagen. At her entry, windows open up once again, allowing in more light and heat. She herself is a heat source, and adds to the energy burden.

The late afternoon brings a dry, hot wind gusting up to 40 miles per hour. This isn't unusual in the desert, and it actually aids the cooling of the heat sink, but it puts pressure on the walls and roof, which are stiffened to compensate. The strong magnetic fields associated with this process could be dangerous to Mom—she's wearing a metal wristwatch and fiddling with a stapler—but the fields are deflected outside by sheets of superconducting material, cooled by still more Peltier junctions. By the time Dad and Sonny return at 5:30, the house's entire east face is one big Peltier heat sink, fiercely hot and color-coded for danger with red and black stripes.

The family dines in the same glassy, east-facing alcove where they breakfasted, and finally, as the sun begins to set, the house begins opening windows on the west side as well. The sunset is beautiful, and before it's finished, the interior lights begin, softly at first, to come on. Once it's fully dark outside, the windows disappear again—there's nothing to see out there in the dark, and even the best privacy glass would leak visible shadows and silhouettes. Then, at 9 P.M., the lights begin to dim. This is the family's cue to begin getting ready for bed, and when they actually turn in at 9:30, the lights go off altogether, and the ceiling turns transparent, offering a beautiful view of the stars and Milky Way as the family drops off to sleep. Later, when the moon rises and the temperature of the desert floor drops sharply, the ceiling turns opaque again, and even drinks in a few milliwatts of moonlight, to help offset the energy needs when the heaters and nightlights come on. Soon it will be morning again, and the cycle will start anew.

City Lights

Houses in the country have always needed a degree of self-sufficiency. In towns and especially cities, things are a bit more complex and intertwined. The behavior described above, while reasonable for a single home in an isolated area, would be highly obnoxious in an urban environment. If these principles were carried to extremes, with each building behaving selfishly to minimize its own energy needs and maximize its internal comforts, then every city could become a war zone of glaring mirrors and heat pollution.

In hot weather, every building would be dumping its heat overboard, and reflecting away as much sunlight as possible. Of course, they already do this, but not so efficiently or systematically. Today's skyscrapers vent their hot air up and away, from the roof; tomorrow's may do so from radiant heat sinks located in deep shadow, occurring mainly at street level. The blackness of the heat sinks would deepen these shadows, and the infrared they emit would mean that, for passing pedestrians, the shade would offer no relief from the heat.

In cold weather, these same buildings would reflect nothing; they'd be pillars of absolute blackness, drinking in every photon they possibly could (except maybe in the deep shadows, where mirrored skin may provide better insulation). The winter streets of such a city would be far colder and darker than the ones of today, robbed of light but not of wind and snow. And the magneto-rheological and electro-rheological substances that stiffen the buildings against these wind and snow loads could result in an entirely new form of pollution: stray magnetic and electric fields of very high intensity. These are not only dangerous in their own right, but also have the potential to interfere with radio signals and nearby electronic devices—possibly including the wellstone skins of neighboring buildings, which would then be obliged to grow superconducting Meissner shields of their own, reflecting the fields back into the street again. This would also have the effect of pushing adjacent buildings apart, resulting in a literal wrestling match where one could potentially damage the other, or force its "surrender" to the invading fields.

At night, the buildings could potentially light up, as present-day skyscrapers do, but they could just as easily remain black, drinking in the glow of streetlights for their own sustenance, and making the night darker. Or perhaps they'd shiny themselves up to retain heat, and turn the streets into weird mirror-mazes of reflected and re-reflected light.

Either way, the effects of such selfishness are far from aesthetic. Every extreme would be amplified: the city's cold, dark corners would be colder still, and black as velvet all around. Bright areas would suffer the additional glare of a dozen new suns, reflected in smooth mirrors rising hundreds of meters above the street. As in a forest, taller buildings with the best access to sunlight would have a huge advantage over the smaller structures living in their shadows. This could well result in an arms race, with taller and taller buildings casting the street below into ever-deeper gloom and heat and field-pollution.

There is also the very real danger of malicious hackers, who would invariably find ways to override or subvert the normal programming of any system. In the end, buildings, like the people inside them, may require sophisticated immune systems and some stiff rules of etiquette in order to get along. The possibilities may be limitless, but in a programmable city, the covenants are likely to be strict. Perhaps there'll be a few holdouts against this ethical tide, just as there are places in America today that resist the alleged "greater good" of zoning laws. Houston, with random skyscrapers jutting up from quiet suburban neighborhoods, is a notably quirky example. In fact, the quirks of virtually any turn-of-the-millennium city could be weirdly enhanced by the advent of programmable matter. Imagine a *genuine* City of Lights or Silicon Valley or riotous Mardi Gras full of levitating magnetic clowns. You heard it here first: the future is darker and brighter and weirder than any cartoon. When technology looks like magic, the world itself becomes a fairy tale.

8

The Future Tense

I ask you to look both ways. For the road to the knowledge of the stars is through the atom, and important knowledge of the atom has been reached through the stars.

—Sir Arthur Stanley Eddington, *Stars and Atoms* (1928)

FOR BETTER OR WORSE, the speculations in this book are mainly my own. When pressed for visions of the future, university scientists tend to clam up, or talk airily about the differences between basic and applied research, or science and engineering. "Scientists," Charlie Marcus observes dryly, "don't necessarily make better futurists than the people down at the donut shop. So much advancement depends on serendipity—the laser came out of microwave research, not a desire by Schawlow and Townes to improve surgery or record players." When goaded with food and caffeine, he does mention the possibility of quantum "spin pumps" and "spin filters" with possible telecom applications. Also weirder stuff like magnetism pumps, although he has no real idea what devices like that might be used for. "Quantum computing," he says dutifully, since that is ultimately where his funding comes from.

Kastner is more direct: he pulls out a graph of federal research expenditures from 1970 to 2000, which shows Life Science unambiguously eating the lunches of Engineering and Physical Science. "The nation is not willing any longer to invest in basic research in the physical sciences, and I would argue that condensed matter physics has been treated worst of all. This is ironic because it has in the past and promises in the future to be the most important field for new technology."

The situation improved somewhat in 2000, when Bill Clinton's National Nanotechnology Initiative allocated $422 million for nano- and mesoscale research for fiscal year 2001. As of this writing, funding was expected to increase to over $500 million for 2002, although the unwelcome intrusion of war presents a new and quite daunting financial challenge. Warily, Kastner concludes, "All this makes us nervous about making claims that could be attacked, even if they are right."

In some sense it's a disservice to belabor this point: there's a bright gleaming future ahead of us, all right, but making promises about it is not in the physicist's job description. "It's easy to raise expectations," Ashoori warns. "It's easy to promise a lot. Applications will happen, but we don't know how or when they'll happen. We still haven't found that killer app."

About the materials-science applications for quantum dots, Quantum Dot Corporation's forty-four-year-old Joel Martin—a Ph.D. chemist as well as an MBA venture capitalist—waxes enthusiastic. "I'm actually working on one right now. This nanoscale technology just gives us another degree of control over the properties of matter, of surfaces, that we don't have by other means. That's incredibly useful."

Alas, too much money can shut people up as surely as too little; Martin is a Valley insider who has participated in nine startups, two of them his own. His board of directors is crowded with other venture capitalists, who eat nondisclosure agreements for breakfast and expect a return on their $40 million investment. I don't bother asking what he's working on—if it pans out, we'll know soon enough. But when I ask his opinion on the prospects for programmable matter in the abstract, he has a one-word answer: "Absolutely."

Still, when it comes to describing the future these devices promise, even Sun's Howard Davidson—rather wild-eyed by the standards of the field—seems to draw something of a blank. That sort of gonzo extrapolation is, properly, the job of engineers and science fiction writers, so the rest of this chapter will follow the path of my own speculations in this area. If programmable matter is indeed a practical goal, there is scarcely a field of human endeavor that won't benefit in some way.

Transportation

The most powerful applications in transportation, as in architecture, have to do with the management of energy. Gathering solar power becomes relatively easy, and it may be possible to create capacitors of variable size to store that energy, and also energy recovered electromagnetically from the braking system and from Peltier junctions harvesting waste heat. The days of the internal combustion engine are surely numbered, and it's conceivable that the hyperefficient electric vehicles that follow may not need to refuel or rejuice or plug into a commercial power supply at all. More likely, they'll take a rapid charge from the home supply, more like a bolt of lightning than a trickle of current, and will run for hundreds or thousands of miles on it.

There's no reason why the street itself couldn't be an enormous solar collector, feeding power on demand up through the car's tires, which could briefly become conductive. If these future tires also incorporate magneto- and electro-rheological capabilities, they may bear very little resemblance to the dainty, air-filled donuts of today's vehicles. More importantly, MR/ER materials in the vehicle frame itself may create crashproof car bodies that limply absorb and distribute the energy of impact, deflecting all harm away from the driver and passengers and then springing back immediately to vehicle shape. Forget airbags; we're talking smooshy-soft dashboards and windshields.

The weirdest transportation innovations may be the ones involving the Meissner effect. If the solar streets were also magnetized, vehicle undersides plated with superconductor could quite easily levitate above them, with no tires at all and no direct expenditure of energy. The superconductor's properties could then be modulated to produce the same sort of propulsive effect used in railguns and experimental "maglev" trains. Stopping and cornering would be more of a challenge, but perhaps we'll always use tires for that. Or feet; once you're levitating, there's no particular reason to have an enclosed vehicle at all, when a superconducting chair or broomstick would work just as well. This isn't an idle speculation; the Meissner effect has been used many times to lift human beings, with no apparent harm.

In fact, it's possible to levitate a living creature without resorting to superconductors at all. Diamagnetism or "molecular magnetism" or "antimagnetism"—millions of times weaker than ferromagnetism—is present in virtually every material, and is particularly pronounced in materials where other magnetic effects are weak or absent. In the 1990s, researchers at Amsterdam's University of Nijmegen demonstrated that a small frog could be levitated in a 10-tesla field, which is comparable to the strength of MRI scanners used in medicine. The frog suffers no ill effects, and since it's being lifted by its very molecules rather than by a platform, it experiences exactly the sort of weightlessness it would in outer space. Contrary to popular belief, though, there aren't "zero gravity chambers" for astronaut training. Levitating a human would require a field strength of 40T and about a gigawatt (1 billion watts) of continuous electrical power. Hardly practical with today's technology. But with tomorrow's. . . .

Even so, while high magnetic fields aren't directly harmful to the human body, they present a major projectile hazard if there are ferromagnetic or even paramagnetic materials lying around. But there is a third form of flight—diamagnetically stabilized levitation—that uses plates of graphite or other diamagnetic materials to tame parallel magnetic fields, making it possible to levitate ferromagnets weighing up to several tons without the need for superconductors or powerful magnetic fields. This does require a second permanent magnet to be located above the levitating magnet, though, so it's more useful for subway trains and frictionless bearings than for automobiles.

Anyway, the magnetic quirks of quantum dots, quantum wells, and programmable matter will likely find their way into transportation applications. Flying cars—a cherished American dream—are definitely possible. Magnetic levitation may also be harnessed to prevent crashes altogether; when collision is imminent, both vehicles can set up slippery, squooshy, mutually repulsive fields, and slither past one another like greased Nerf balls.

Arts and Fashion

The phenomenon of "color" is much richer than the visible spectrum, and occurs largely within the visual processing centers of our brains. The spectral colors—what we might call "emissive colors"—don't account for factors other than wavelengths of light, but our eyes and brains are capable of interpreting extremely subtle gradations of texture and intensity, absorption and refraction. For this reason, it's quite possible to invent new "colors" and even whole classes of colors. People do it all the time, and so does nature. Dyed silk has a completely different appearance than cotton or wool or flax linen colored with the exact same dye. Blue feathers look very different from blue sky or blue berries or blue-dyed aluminized mylar.

Fluorescent colors are tinged with whiteners that reflect disproportionate amounts of blue and also absorb UV light and re-emit it as blue or violet, making these colors appear brighter than the objects around them. In fact, under strong UV light they actually appear brighter than the ambient visible light should allow. They fluoresce. And of course there are colors that actually glow, as in backlit glass or the phosphors of a TV screen. The metal-flake paint jobs applied to cars provide an illusion of depth, while holograms carry this property to an extreme, creating entire three-dimensional images "beneath" (or in some cases "above") the surface to which they're applied. And there are holograms and paints that give off a completely different appearance depending on viewing angle.

Quantum dots will probably provide us not only with new sorts of colors we've never seen before but with the power to change these colors under a variety of different influences. Wellstone art could include features too fine for our eyes to discern, or (more interestingly) features at the very edge of our perception. The rainbow sheen of oil on water is an example of visible mesoscale structure; countless others probably exist, and I suspect many of them will induce headaches or even epileptic seizures on sufficiently close inspection.

A hologram is essentially a photograph of interference patterns, generated when laser light reflects from an object and interacts with itself. In effect, it's a photograph of the light waves around an object rather than of the object itself. This occurs at single frequencies; color holography is simply three different holograms, in three different spectral colors, pressed onto the same photographic medium. Realistic holograms can be impressed on films with grain size ranging from 100 nanometers to several microns, but once they're exposed, the individual grains become fixed pieces of a fixed image. A "moving hologram" today simply consists of multiple static images developed into the photograph at different angles.

Sheets of a wellstone-like material, though, would have programmable "grains" only tens of nanometers in size, and could in principle be used to capture, play back, and even transmit holographic images. Moving, color images. A wellstone TV might look less like a moving picture and more like a window into a real, three-dimensional space. You could paste these all over the inside of your house, making it look (from the inside) as though you were living in Tahiti, or Antarctica, or on the red-hot surface of the planet Mercury.

Of course, you could make an ordinary flat TV screen as well, or even an animated "painting" or "mosaic" or "fresco" whose apparent optical/chemical/physical properties resembled paint or ceramic tile but changed on demand to make the image move. Imagine picking an "acrylic on drywall" theme for your computer desktop and having it really look that way! Wellstone also offers the possibility of transmitting nonvisual stimuli, such as temperature, flavors, and tactile sensations like firmness and thermal conductivity. And the pain of electric shock, sure. A virtual reality based on programmable matter would be vastly more realistic than one based on present-day audio, video, and "haptic" (tactile) interfaces.

The same effects could be applied, weirdly, to clothing. With its tight weave and mesoscopically fine fibers, wellstone itself would make a poor textile, resembling plastic sheeting more than cloth. But wellstone fibers could be interwoven with other textiles to make them programmable. This concept goes "wearable computers" one better, and opens the door to solar-powered and movement-powered garments that can store whole libraries of music and video and text, keep you cool in the summer with flexible Peltier panels, provide headlights and taillights in the dark, and even protect you from certain types of injury through a combination of MR, ER, and maglev techniques. Even without the benefit of wellstone, a Brussels-based consortium called Starlab is already producing "electrically aware" jackets and other garments, with power, sensor, and computing elements stitched right in. This may sound far-fetched, but in fact Starlab's owners and funding sources include clothier Levi-Strauss, electronics giant Phillips, shoe maker Addidas, and luggage kingpin Samsonite. All are sober, profit-minded companies that expect to make money off this venture sooner rather than later. Other players in this arena include IBM, an MIT spinoff called Charmed Technology, and wearable computer manufacturer Xybernaut.

Potentially, your smartjacket can even be friends with you, through the miracle of artificial intelligence. Some scientists—most notably physicist Roger Penrose—have speculated that true computer intelligence will never be possible, and that human consciousness relies on subtle quantum effects in the microtubules that form the soft internal skeleton of nerve cells. Penrose argues, essentially, that the human spirit (as opposed to the wiring) is actually a form of quantum cellular automaton. I have grave doubts about this idea for a variety of reasons, but even if it's 100 percent true, wellstone fibers are electrically quite

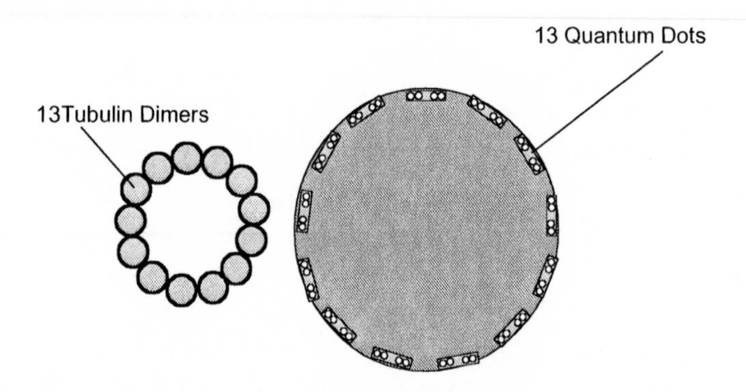

FIGURE 8.1 COMPARISON OF MICROTUBULE AND WELLSTONE FIBER

similar to microtubules. A microtubule is a hollow, 25-nanometer-wide cylindrical stack of protein molecules known as "tubulin dimers," of which there are thirteen in the circumference of the cylinder. Each of these forms a "cell" in the supposed cellular automaton, which runs around the fiber and along its length, passing excess electrons around in complex patterns. (See Figure 8.1.)

Coincidentally, a typical wellstone fiber as envisioned in Chapter 6 of this book is only about twice as thick as a microtubule, and also contains about thirteen "cells" (quantum dots) around its circumference. These should be quite capable of generating quantum effects like the ones observed in microtubules, so when gathered into branching structures analogous to the cytoskeleton of human neurons, they should exhibit much the same behavior. So even if "neural network" computers are never able to mimic human intelligence, wellstone quantum networks would likely be able to.

Industry and Commerce

The principles described above apply to virtually any object in the programmable world. Pseudomaterials could be specified that are photovoltaic enough to keep themselves alive on ambient light, and that have other desirable properties such as embedded computing and piezoelectric microphones for voice command. These materials might resemble natural ones for aesthetic reasons (e.g., decorative wellgold and wellmarble and wellwood), or could have completely unnatural properties for other purposes. You never know when you might need a luminous, diamagnetic superconductor.

Also, despite its solid-state design, wellstone is capable of weakly interacting with other objects. It can grasp atoms and molecules, and even pass them around from one dot to the next. Thermal bombardment is a real problem here, but this can be solved by keeping the wellstone surface at a very low temperature. A liquid nitrogen bath is the most obvious method for this, although with Peltier cooling and a good heat sink, it may be possible for the wellstone to generate its own cryogenic temperatures. Either way, a sort of two-dimensional "bucket brigade" could conceivably be set up to sort and manipulate molecules and small particles, and perhaps even assemble them into larger structures.

This is very unlikely to be more efficient than traditional chemical or industrial manufacturing. Then again, if we have wellstone ceilings and floors sitting around idle, this sort of thing—however inefficient—could be a "screen saver" activity, perhaps oriented toward dusting and disinfecting a room when no one is around, and transporting the offending particles to a waste container or incinerator. Alternatively, the same process could remove impurities from the air, or even assemble scent molecules to sweeten it. Disassembly of macroscopic garbage this way is probably impossible, not to mention unnecessary when cruder techniques such as laser incineration would be readily available.

More practically, wellstone's ability to modify its surface properties might result in a variety of novel "coatings" to repel water or paint, attract pollutants, resist corrosion, and even serve as programmable catalysts. In chemistry, a catalyst is any substance that increases the rate of a desired chemical reaction without itself being consumed. For example, iron is used to catalyze the production of ammonia, while aluminum chloride is used to separate butane molecules. Certain molecules, known as "polyfunctional" catalysts, have multiple active sites or are capable of changing shape or subtly altering their electrochemical properties, so that they can catalyze more than one reaction in a series. Polyfunctional catalysts are widely used in the petroleum industry, for example, to enable multiple steps in the crude oil "cracking" process. This saves time and money over the alternative of adding new catalysts at each step. The potential of quantum dots in polyfunctional catalysts is definitely worth exploring.

Medicine and Biology

Biological catalysts, known as enzymes, regulate virtually every chemical reaction that takes place in a living cell. More than a thousand different enzymes have been identified, each controlling one specific piece of the great puzzle that is life. Without enzymes, most biological reactions would occur much more slowly or not at all. The vast majority of enzymes are proteins, which are linear polymers composed of amino acid building blocks. Proteins fold into complex three-dimensional shapes whose geometry and charge distribution dictate their function, just as the shape of a comb or shovel or key dictates how it will interact with the world. Proteins are also flexible, and many of their functions rely on their ability to change shape, even sometimes to act as tiny machines.

Wellstone fibers will be much too thick and rigid to simulate proteins, but then again most enzymes resemble large tangles of string. They include a lot of extraneous and structural material, and rely for their functioning on only a handful of active sites. Since buckyball-sized quantum dots can generate bonding forces approaching those of biological receptors, complex assemblies of them could conceivably reproduce certain aspects of these active sites, and thus yield a weak but programmable enzyme.

Larger devices, either of wellstone or of other quantum dot materials, could serve in the human body as microscopic detectors of medically significant phenomena such as heat, light, pressure, electric fields, and chemicals. Optical or magnetic quantum dots, bonded to enzymes and other molecules, could even reside permanently in our cells, enabling entirely new diagnostic techniques. Light-sensitive quantum dot surfaces may serve as "cameras" for the microscopic exploration of the body, and their ability to generate laser light at a variety of frequencies may make them invaluable in microsurgery. Precisely tuned lasers could break up artereosclerotic plaques or blood clots while leaving healthy tissues intact.

Programmable matter applications need not be restricted to the curative. A thin disc of wellstone inside the eye—or a contact lens on the eye's surface—could photodarken to protect the retina against bright lights (even extremely sudden ones). It could also act as a permanent "head-up display," projecting pre-lensed holographic information onto the retina. Piezoelectric wellstone in the ear could serve as both a receiver and a generator of sound waves.

There may also be internal applications for variable stiffness. Sexual jokes aside, it's quite conceivable to supplement the human skin or skeleton with MR/ER materials that could dramatically improve our durability. To get really far out for a moment, it's even conceivable we could supplement our slow, classical, chemically based nervous systems with superfast, massively parallel quantum hypercomputers. Early experiments have already bonded living neurons with silicon chips, and successfully transferred signals between them, so if we find *any* reason to put wellstone in our bodies, we may get this benefit as a freebie.

Military and Aerospace

Future soldiers could benefit from most of the same architectural, vehicular, clothing, and biomedical applications already discussed. Any technology that makes people safer, more comfortable, or better informed will always go over well in military circles. Armies and navies do of course have the additional requirement of camouflage—the less visible soldiers and equipment can be, the greater their security and element of surprise. This is why we have stealth aircraft, which absorb radar waves rather than reflecting them. In color terms, they are radar-black.

If transparent wellstone were able to take on a very low index of refraction (close to 1.0), its optical characteristics would resemble those of air or vacuum. This is *true* invisibility, and it could result in aircraft that did not show up on any electromagnetic scanners of any kind, including the human eye. Boats and submarines would be even easier, as the refractive index of water (1.3) is only slightly less than that of glass (1.5–1.7). Unfortunately, this would not protect the crew inside the vessel from being seen. Still, it's possible to achieve "cheap invisibility" through simple video technology: make the entire surface of the vehicle both a video camera and a video projector. Every part of the exterior displays an image of the light shining on the opposite side, including full visual reproduction of grass, sky, rocks, buildings, and whatever else is around.

Britain has experimented with electro-optical camouflage since World War II, when it placed floodlights on the wings of antisubmarine aircraft to blend them into the sky. Today in the United States, there are strong indications that both industry and government are pursuing this technology in earnest. It's estimated that an M-1 battle tank could be rendered "invisible" for about $10 million. The illusion would be far from perfect—a moving tank would look like a boxy distortion in the air, especially if its video screens were dirty or damaged. But it would be enormously better than the simple paint splatters we use today. And with the capability to display holographic images as well as flat ones, programmable matter could possibly make the camouflage even more convincing.

Defeating sonar would be more difficult, but naval submarines today are coated with vibration-dampening (acoustically black) materials. Wellstone may have some limited ability to behave this way. For active suppression, piezoelectric materials that detect vibration, and generate "canceling" waves 180° out of phase under the control of various computer algorithms, are already components of noise-canceling systems for hearing protectors, helicopter engines, and audio equipment. It's conceivable that sound-canceling equipment could be incorporated into vehicles to suppress constant, repetitive, or slowly changing sounds. (For what it's worth, this idea was the basis for my first science fiction story, published in 1987.) Whether this capability is worth the expense and effort of dedicated hardware is difficult to say, but if your vehicle is already covered in programmable matter, acoustic dampening or canceling becomes more a problem of software than of hardware.

One of the weirdest applications is being investigated by physicists at Positronics Research and the Air Force Research Laboratories (AFRL), who are hoping to use quantum dot materials for the long-term storage of positronium. This is a form of "atom" consisting of one electron and one positron, or anti-electron. Antimatter, yes, just like on Star Trek. But this is not science fiction: according to their calculations, a man-portable brick of such material could be capable of storing up to 10 μg (0.000001 grams) of positronium—equivalent to over 900 pounds (408 kg) of a high explosive such as TNT. That's a hell of a grenade by anyone's standards—imagine a BB gun that can sink a battleship—yet it's also "safe" in the sense that this particular matter-antimatter reaction doesn't release large amounts of harmful radiation. And going boom isn't the only thing positronium can do; AFRL has already commissioned detailed designs for an antimatter-powered jet aircraft that should be capable of traveling for months without refueling. With minor tweaks and additions, the vehicle would probably be capable of reaching orbit.

Programmable matter could also offer dramatic improvements in some of the most difficult aspects of spacecraft management: solar power generation, efficient heat absorption and radiative cooling, and communications. When your entire hull can be reconfigured into various flavors of conformal antenna, even extremely noisy or low-powered signals will be able to get through, and high-powered, high-frequency signals will be capable of much higher throughput (bits per second) than is possible with current systems. And with the vast computing power that will almost certainly be available, new forms of data compression and error checking may improve performance still further.

When briefed on the concepts of programmable matter, Robert M. Zubrin—president of the Mars Society and CEO of Pioneer Astronautics—became an immediate proponent. Zubrin has spent considerable time in simulated space habitats, and is well aware of the problems astronauts face. "This would be marvelous," he says. "It would solve all our thermal problems with spacesuits and spacecraft. And in a hab module, being able to place a window anywhere you want is very desirable psychologically. You could also emit UV light to sterilize things, including the air and plumbing.

"And think of the science: if I had a little card I could look through, that could flip through a series of different optical filters, I could tell what the rocks were made of by how they reflect the light. Heck, I could integrate the spectral analysis right into it, like a Star Trek tricorder. And that card would be my binoculars and night vision scope, too. I'd store a million books on it. I'd use it as my cell phone."

More airily, he notes, "This material would be very useful in solar sails. You could steer a spaceship with changes in reflectivity—a completely solid-state control system. Boy, if I could take in incident light from any direction and release it in a highly directional way, maybe as a laser, that would be a kind of photon drive. Free propulsion like that would be very desirable."

Wellstone's ability to generate laser beams also has clear military value. In fact, it's not difficult to imagine a monolith whose programmable substance is alternately a sensor, a computer, a power source, a communication device, an electro-optical weapon, and a resilient, mirror-bright armor.

This kind of dramatic and instantaneous transformation is serious Clarke's Law mojo—virtually indistinguishable from magic. Can programmable matter do *all* of the things I've outlined in this chapter? Probably not, and even some applications that are possible may not be practical. Still, some or even most of these capabilities are likely to be realized in the coming decades—a prospect that has many people praying for continued good health. It will probably be 2015 or 2020 before the first serious programmable matter chip is built, and another twenty years before anything like bulk wellstone becomes anything like possible. From there, it will take still more time before the technology becomes widely available.

But when it does—even if it's only a fraction as capable as I've hinted here—its transformative effects will be staggering, a tumble through the rabbit hole for all of us.

If You Thought That Was Crazy . . .

The hacking of matter is by no means restricted to quantum dots. Microelectromechanical systems (MEMS) technology uses the familiar techniques of microelectronics to produce *mechanical* chips, including micron-sized gears and motors, highly sophisticated pumps, locking mechanisms, manipulator arms and "fingers," and even complete working machines such as heat-driven steam engines. MEMS suffer from a number of reliability problems, mainly because they use silicon instead of more mechanically inclined materials. It's easier; we already know how to work with silicon on the microscale, whereas other materials remain largely unexplored. Still, MEMS are already a hot area of both research and commerce, with a growing number of real-world applications.

The related field of microchemistry uses similar techniques to construct "labs-on-a-chip" that mix and match chemicals in tiny but very precise proportions. These systems are expected to play a major role in the detection of hazardous materials, including bacteria, viruses, and biowarfare agents. There may even soon be factories-on-a-chip that take in air, tap water, and electricity and produce useful chemicals such as soap or sugar.

In the early 1990s, J. Storrs Hall of Rutgers University proposed a particularly ambitious application for MEMS technology: a "utility fog" of twelve-armed, dust-sized silicon micromachines. (See Figure 8.2.) These "foglets" would be capable of joining hands in a variety of configurations, as well as extending and retracting their arms, and adjusting the optical characteristics of their faces. En masse, they create a programmable substance that can, on command, change both its shape and its density.

As an animate modeling clay, utility fog offers a number of fascinating capabilities that quantum dots do not. Physically speaking, the technology doesn't seem too far out of reach, although the software and power distribution challenges are formidable. Of course, foglets can't reproduce or build anything else, and their texture and temperature and glassy composition would prevent them from masquerading as any other material. Nor would they make good computers, mirrors, power cables, or floors. They're by no means a technological panacea, but they do fill in some gaps left open by other technologies.

These "top-down" approaches are neatly complemented by the "bottom-up" techniques of chemistry, nanoelectronics, and molecular nanotechnology. All three fields are seeking, through different channels, to produce intricate designer molecules that perform precise electrical and mechanical tasks at the nanoscale. In combination, these molecular machines could form laboratories and factories a thousand times smaller than the MEMS versions discussed above, and might even lead to semi-autonomous manufacturing nanorobots capable of producing an unimaginable variety of goods, including exact copies of themselves. Nanotech visionary K. Eric Drexler has coined the widely adopted term "assemblers" to describe such devices. The feasibility of assemblers is demonstrated by, among other things, the existence of life, which after all produces copies of itself in addition to various other products, including the oxygen in our atmosphere. It remains to be seen whether human beings can build assemblers for our own strange purposes and, if so, whether we can avoid the obvious acci-

FIGURE 8.2 UTILITY FOGLET

This microscopic machine can extend or retract its twelve arms on demand. When connected to millions of similar machines, it creates a "utility fog" of smart matter capable of changing its shape, density, texture, and color. (Image courtesy of J. Storrs Hall.)

dents of runaway replication. But I very strongly suspect that the answer to both questions is yes.

This of course has nothing to do with quantum dots or artificial atoms. My point is that no technology, however wonderful, exists in a vacuum. By the time a wellstone-like material becomes widely available, there will have been stunning advances in these other fields as well. It's quite likely that the most useful products will blend many of these functions together. There will almost certainly be designer molecules that also manipulate electrons in clever ways, and there's no reason why a wellstone fiber or colloidal quantum dot solid couldn't also incorporate molecular machinery.

And the fun doesn't have to end there. A December 1993 discussion between Hugo de Garis and J. Storrs Hall, on the Usenet newsgroup *sci.nanotech*, included speculation about possible "picotechnology," taking place on the scale of atomic nuclei, and "femtotechnology," taking place on the scale of quarks. De Garis mentioned that Drexler was skeptical when presented with these ideas, and Hall replied that such technology probably could not operate outside extreme environments, such as the crushing surface of a neutron star. Still, in his 1998 novel *Diaspora*, science fiction writer Greg Egan took a playful look at short-lived "halo nuclei"—atoms on the edge of nuclear stability, in which some protons or neutrons have wandered away from the nucleus to surround it in a cloud with some similarities to the orbital structure of electrons. Egan suggested that such atoms, under proper inducement, might briefly interact in nucleo-chemical ways, inducing useful changes in the matter around them. As we've already seen, crazier notions have turned out to be true.

Nor is this the only suggestion for matter's subatomic rearrangement. Supercooled atoms can be coaxed into giving up their particle nature and behaving together as a single macroscopic wave, in a state of matter known as a "superatom" or Bose-Einstein condensate. These materials have *extremely* weird properties, and have been used to slow down light beams to a few meters per second, or even (like the praesodymium yttrium silicate material described in Chapter 3) to stop them altogether. That's a refractive index of infinity—large by anyone's standards. This principle may someday be harnessed on computer chips, to produce light-storing components that would otherwise be impossible.

In 1997, MIT researchers even succeeded in harnessing the wave properties of these condensates to create "atom lasers," which transmit mass-energy in much the same way that laser beams transmit light energy. Atom lasers don't travel at the speed of light, and as a consequence, the "beam" may droop noticeably with gravity, like the stream of water from a firehose. Still, remarkable progress has been made in manipulating these beams with crude "optics," including atom mirrors, atom beam splitters, and atom diffraction gratings. Many scientists, including Lute Maleki of NASA's Jet Propulsion Laboratory and Pierre Meystre of the University of Arizona, have speculated that just as lasers can be used to create the three-dimensional images we call holograms, atom lasers may someday be used to create "atom holograms," which would in effect be atomically precise duplicates of the object being imaged. Star Trek's imaginary "replicator" and "transporter" technologies spring immediately to mind.

I'm personally dubious about this—it's not clear how anything other than the outermost surface of an object could be "imaged" and reproduced in this way—but the idea does point to one more method by which a slice of matter, or at least its surface, might be dramatically modified in real time. This could prove to be an extremely efficient, precise, and repeatable way to "print" nanoelectronics onto a chip, or to build up the layers of a nanoscale fiber all at once. *That* would sure solve some manufacturing bottlenecks. In fact, if nanometer-scale circuit traces could be laid down, erased, or modified in this manner, it might even be possible to alter the dimensions or design of quantum dots in mid-operation.

Finally, we have the possibility of "matter" that isn't there at all. Physicist Don Eigler is famous for being the first person to assemble a recognizable object by picking and placing individual atoms. In 1989, he spelled "IBM" in nanometer-sized letters using xenon atoms on a sheet of glass. More recently, Eigler has investigated "quantum mirages," in which a magnetic atom placed at one focus of an elliptical quantum corral produces the quite realistic mirage of an atom at the other focus. Thus, a single atom can effectively be in two places at the same time. Practical uses are difficult to imagine, but as science progresses from hacking matter to actually engineering it at the subatomic level, such phenomena may become commonplace.

Undoubtedly, there are hard physical limits on what human beings can accomplish through the tweaking of matter's building blocks. But I, for one, find it exciting and reassuring that we're nowhere near these limits. In fact, we can't even guess where they are. Until we hear differently, the major restriction is our own feeble imagination.

AFTERWORD: ACCIDENTAL DEMIGODS?

Gathered together at last under the leadership of man . . . unified, disciplined, armed with the secret powers of the atom and with knowledge as yet beyond dreaming, Life, forever dying to be born afresh, forever young and eager, will presently stand upon this earth as upon a footstool, and stretch out its realm amidst the stars.

—H. G. Wells, *The Outline of History* (1920)

ONE FLAW WITH CLARKE'S LAW is really an ambiguity in the English language: the exact definition of "magic." Is it alchemy, or divination? Is it fairies buzzing around a toadstool? Or is it something deeper: the raw, mythic power of a Camelot, or even an Olympus?

Vernor Vinge, a professor of mathematics and computer science at San Diego State University, has written extensively on the subject of exponential technological growth. For several decades now, he's been fascinated by observations such as Moore's Law, and the steady expansion of prices and stock markets and the human population. But above all, he argues, it's the explosion of human knowledge that promises to thrust us into some new and incomprehensible era.

In Greek and Roman times, an educated person was expected to know everything—literally—and this was an achievable goal since even a modest library could contain every important fact and idea that had ever been committed to writing. Plato and Aristotle conspired to create the first encyclopedia, a single document that summarized this knowledge for the benefit of future scholars and philosophers. But by the Middle Ages, a comprehensive encyclopedia sprawled across eighty separate volumes, and by the end of the eighteenth century so much was known that it was considered impossible to summarize all of it.

Since that time, we have evolved the idea that centers of learning—libraries, universities, and so on—are places beyond human scale. We journey through them, as through a world, and we never dream of knowing or seeing everything they contain. What we value is the *trajectory* of the journey, and what we hope is that our velocity is sufficient to carry us beyond some particular boundary of human knowledge. It's a race against time, against death: with so much already known and studied and catalogued, can we learn something genuinely new in the short span of a human lifetime? Can we add a line or a page to the human library? It seems a daunting task indeed, and yet ten thousand new books are published every month worldwide, along with articles in thousands of different trade and technical journals. Some experts estimate that the sum total of human knowledge now doubles every four years.

Even the Library of Congress has stopped trying to keep up with all of it, although the astonishing growth of the World Wide Web—and the search engines that index it—may soon restore the capability to have all of human knowledge at our fingertips. And with increasingly sophisticated artificial intelligence, there will be "autonomous agents" capable of gathering and summarizing information for our own individual purposes—and perhaps *their* own purposes as well.

Vinge's take on all this exponential growth is that it leads, inevitably, to a point where all the growth curves go vertical, where all of our assumptions break down—in mathematical terms, a "singularity." On the far side of this event, the style and substance of human life will have undergone a transformation so dramatic that we can scarcely comprehend it. And for all we know, quantum dots may be the straw that breaks that particular camel's back. If programmable matter can be made to work, not only is it a powerful technology in its own right, providing hypercomputers and dazzling optics and an unimaginable wealth of new materials—it also has an "instant gratification" quality that eliminates many of the inconveniences of today's world. New materials can be invented and examined at any time, without the need to mine or manufacture or calculate anything. New devices can simply be specified and tried—eliminating the pesky steps of design iteration, parts and materials acquisition, prototyping and production. Matter becomes something akin to software—infinitely malleable and precisely obedient. In the future, manufacturing may join mathematics and software as an enterprise where true, literal perfection is not only possible but easy and quick. This makes the technology of programmable matter a huge enabler and catalyst for technology development in other areas.

Perhaps more important, wellstone provides ready access to the quantum world. Without specialized equipment of any kind, even home hobbyists will have the means to explore and manipulate infinitesimal particles and waves and energies, as easily as backyard astronomers examine the heavens today. Maybe nothing significant will come of this, other than a general increase in the understanding of quantum mechanics. It's far more likely, though, that this will drive scientific and technological progress even further and faster than before.

The classic Promethean warnings apply: new technology brings new dangers. In Greek mythology Prometheus was a Titan, or demigod, who stole fire from the forge of the gods and delivered it to human beings. Humans abused the gift, though, falling out of harmony with nature and creating a scientific civilization that rivaled—and therefore offended—the gods. Prometheus was chained to a rock as punishment, and sentenced to be pecked at forever by a patient and sharp-taloned eagle. Those Olympian deities guard their power jealously.

Is the human race collectively ready for point-and-click power over matter itself? Certainly, any abuses that are possible are also very likely to be attempted. Still, as with most other technologies, it seems unlikely that the potential evils of programmable matter outweigh the benefits. Indeed, on the whole our civilization will be greatly improved. There will be much less need to harvest and squander the natural resources of our planet—other than silicon and sunlight, which can be found literally everywhere. Mines and factories will give way to vast software libraries and home workshops. Our newer, cleaner industries will produce superior products at much lower cost, with vastly less waste and pollution. We will have efficient ways to capture and store and reuse energy, possibly including the long-term storage of antimatter.

By the end of the 1950s, most Americans expected the twenty-first century to be a time of conquest and exploration, as humanity spread to the planets, and perhaps even the stars. That dream has faltered, owing mainly to the staggering cost of conventional rockets, which are as large and heavy and complex as office buildings, yet as disposable as paper napkins. Technologically, they've changed very little since the earliest days of the space program. But with programmable matter at our command, we may find it trivial to construct much lighter and more capable spacecraft, even as we wallow in an energy glut of unprecedented proportion. Effectively, we'd all be multibillionaires, with the economic and industrial clout of entire twentieth-century nations at our command.

When we do settle the planets, it may not be as overgrown cavemen, dressed in skins of airtight fabric and hurling metal spears at the sky. Instead, we may stride there as a budding new race of Titans, with the gifts of Prometheus cradled lovingly in our open hands. And from there, my friends, the stars themselves await.

POSTSCRIPT

SINCE THE SUMMER OF 2002 when the final draft of this book was completed, a few noteworthy changes have occurred. Galileo Shipyards LLC, the aerospace R&D company I run with Gary Snyder, has formed a subsidiary called The Programmable Matter Corporation, specifically geared toward wellstone and quantum dot research. Our patent application is still pending, but with new legal representation we're pursuing it aggressively.

More important, a number of low-cost technologies have emerged for the production of electronic devices on the nanoscale, including dip pen nanolithography and nanoimprint lithography. Practical advances have been made in electron beam lithography as well, so that the equipment to produce 10-nm features is within the reach of many small- and medium-sized businesses. Progress in these areas has been amazingly rapid. We're currently in discussion with a number of vendors and are examining alternative wellstone designs that exploit the features and drawbacks of these various systems.

At the same time, simplified techniques have become available for the simulation and modeling of electrons in quantum confinement. An excellent reference is Paul Harrison's *Quantum Wells, Wires, and Dots* (John Wiley & Sons, 2000).

In addition, our work has attracted interest at several government agencies, including NASA and DARPA. The process of applying for research grants can sometimes drag out for years, but near-term funding for this research is under discussion as I write this. We have also been approached by a steady stream of venture capitalists, whom we're mostly turning away until these concepts can be proven out in the laboratory, and their commercial potential is better defined.

At any rate, my Chapter 8 estimate of a 2015–2020 date for the first programmable matter chip now seems foolishly pessimistic. In fact, the technology to produce these may be very nearly at hand, and something similar to the wellstone fibers of Chapter 6 may be feasible before the end of this decade, although they will likely be flat ribbons rather than cylindrical fibers. Their exact properties won't be known until they're available in bulk, however, which could still take a while.

Almost by definition, in the fields of science and technology a printed book is out of date well before it reaches the bookstore shelves. This volume is no exception. However, up-to-the-minute information can be found in my Programmable Matter FAQ, located at *http://www.sciencebar.com/pmfaq.htm.* Readers are also welcome to submit any questions that aren't answered in the FAQ. All best wishes,

Wil McCarthy
December 2003
Lakewood, CO

APPENDIX A: REFERENCES

Chapter 1

Clute, John, and Peter Nicholls. "Clarke, Arthur C." *The Encyclopedia of Science Fiction* (St. Martin's Press, November 1995).

Diamond, Jared. *Guns, Germs, and Steel* (W. W. Norton & Company, 1998).

Encyclopedia Britannica, 2001 Edition: "Condensed Matter Physics," "Clarke, Arthur C.," "Technology," "The Main Theories of Magic," "Superstring Theory."

Freitas, Robert A., Jr. "Navigation" and Appendix A. *Nanomedicine, Vol. 1: Basic Capabilities* (Landes Bioscience, 1999).

Garraty, John A., and Peter Gay. *The Columbia History of the World* (Harper & Row, 1981).

CHAPTER 2

Cho, Adrian. "Nanotubes Hint at Room Temperature Superconductivity," *New Scientist*, November 30, 2001.

Encyclopedia Britannica, 2001 Edition: "Digital Circuits," "Semiconductor Device: Metal Oxide Semiconductor Field Effect Transistor."

Hodgin, Rick C. "Intel Creates 20nm Transistors," *Geek.com* (June 11, 2001).

Kastner, Marc A. "Artificial Atoms," *Physics Today*, January 1993.

______. Personal interview, July 13, 2001.

______. Personal communications, June 2001–September 2002.

Kogan, Andrei (MIT). Personal interview, July 13, 2001.

Kouwenhoven, Leo, and Charles Marcus. "Quantum Dots," *Physics World*, June 1998.

Likharev, K. K. "Single-Electron Transistors: Electrostatic Analogs of the DC Squids," *IEEE Transactions on Magnetics*, March 1987.

McCarthy, Wil. "Quantum Computers: The Secret Is Out," *Science Fiction Weekly*, August 2000.

Peterson, Ivars. "Electrons in Boxes," *Science News*, April 11, 1998.

Turton, Richard. *The Quantum Dot: A Journey into the Future of Microelectronics* (Oxford University Press, 1995).

CHAPTER 3

C. W. "Quantum Dots Stack into a 3D Array," *Science News*, December 12, 1998.

American Association for the Advancement of Science. "Trends in Federal Research by Discipline, FY 1970–2000," 2001.

Ball, Philip. "Solid Stops Light," *Nature*, January 10, 2002.

Bawendi, Moungi (MIT). Personal interview, July 2001.

Bertness, K. A., et al. "29.5%-Efficient GaInP/GaAs Tandem Solar Cells," *Applied Physics Letters*, August 22, 1994.

Byrne, Peter. "Small Wonders," *San Francisco Weekly*, May 18, 2001.

Davidson, Howard (Sun Microsystems). Personal interview, June 15, 2001.

Encyclopedia Britannica, 2001 Edition: "Optical Ceramics," "Industrial Glass," "van der Waals Bonds," "Colloid," "Significance of the Structure of Liquid Water," "Photoelectric Effect," "Luminescent Materials and Phosphor Chemistry," "Clay Mineral: Pyrophylite Talc Group."

Freitas, Robert A., Jr. Appendix A. *Nanomedicine, Vol. 1: Basic Capabilities* (Landes Bioscience, 1999).

Graham, James (UC Berkeley). Personal communication, July 2001.

Graham-Rowe, Duncan. "Cool Colours, Man," *New Scientist*, Vol. 21, April 2001.

Landis, Geoffrey A. Personal communication, August 15, 2002.

Leatherdale, C. A., C. R. Kagan, N. Y. Morgan, S. A. Empedocles, M. A. Kastner, and M. G. Bawendi. "Photoconductivity in CdSe Quantum Dot Solids," *Physical Review B*, July 15, 2000.

Levine, Judah. "Physics 1230: Light and Color," University of Colorado Department of Physics, *www.colorado.edu/physics/phys1230* (Fall 2001).

Martin, Joel (QDC). Personal interview, June 8, 2001.

Nirmal, N., B. O. Dabbousi, M. G. Bawendi, J. J. Macklin, J. K. Trautman, T. D. Harris, and L. E. Brus. "Fluorescence Intermittency in Single Cadmium Selenide Nanocrystals," *Nature*, Vol. 383, 802 (1996).

Noll, Stephen J. "Amateur Lightwave Communication. . .Practical and Affordable," Microwave Update conference proceedings, 1994.

Phillips, C. C., B. Serapiglia, and K. L. Vodopyanov. "Intersubband Spectroscopy of III-V Quantum Wells," *Experimental Solid State Physics*, February 6, 2001.

Quantum Dot Corporation home page, *www.qdots.com* (July–October 2001).

Sacks, Oliver. *Island of the Color Blind* (Vintage Books, January 1998).

Weiss, Peter. "Light Comes to Halt Again—in a Solid," *Science News*, February 9, 2002.

______. "Anatomy of a Lightning Ball," *Science News*, February 9, 2002.

Author's note: Moungi Bawendi has stated that the text of this chapter, though reviewed extensively by outside sources, contains several "errors of fact" whose nature he did not specify. Although independent corroboration has been sought wherever possible, parts of this chapter have an air of unauthorized biography, and should be taken with a grain of salt.

CHAPTER 4

Bigelow, Stephen. *Troubleshooting, Maintaining, and Repairing PCs* (McGraw-Hill, 1998).

Encyclopedia Britannica, 2001 Edition: "Boron Nitride," "Helium," "Capacitance," "Analysis of a Thermoelectric Device," "Heat Flow," "Conducting Properties of Semiconductors," "Piezoelectricity."

Fairley, Peter. "Ultrahybrid," *Technology Review*, September 5, 2001.

Freitas, Robert A., Jr. "Pathways to Molecular Manufacturing," "Power," and Appendix A. *Nanomedicine, Vol. 1: Basic Capabilities* (Landes Bioscience, 1999).

Glover, Thomas J. *Pocket Ref* (Sequoia Publishing, 1989).

Horn, R., R. Neusinger, M. Meister, J. Hetfleisch, R. Caps, and J. Fricke. "Switchable Thermal Insulation Applied in Building Facades—Results of Computer Simulations," Bavarian Center for Applied Energy Research (ZAE Bayern), July 5, 2001.

Johnson, A. T. ("Charlie"), Jr. (U Penn). Personal interview, October 2002.

Kanellos, Michael. "IBM Chips Bulk Up on Carbon Nanotubes," ZDNet News, *http://zdnet.com.com/2100–1103–917412.html* (May 20, 2002).

Lieber, Charles M. "The Incredible Shrinking Circuit," *Scientific American*, September 2001.

Marcus, Charles (Harvard U.). Personal interview, July 16, 2001.

______. Personal communications, July 2001–September 2002.

McCarthy, Wil. "Proposal for a Wearable Operator Control Unit," Omnitech Robotics Inc., 2000.

______. "Ultimate Alchemy," *Wired*, October 2001.

______. "A New Kind of Cool," *Wired*, March 2002.

Nippon Graphite Fiber corporate web site, *http://plaza6.mbn.or.jp/~NGF/english/*, July 2001.

Peterson, Ivars. "Polyhedron Man: Spreading the Word About the Wonders of Crystal-Like Geometric Forms," *Science News*, December 22, 2001.

Roukes, Michael. "Plenty of Room, Indeed." *Scientific American*, September 2001.

Thompson, Howard. "Economics of Large Space Projects." University of Wisconsin NEEP 602 Lecture #37, *http://silver.neep.wisc.edu/~neep602/lecture37.html*, April 24, 1996.

Venkatasubriaman, Rama (RTI). Personal interview, November 29, 2001.

Weiss, Peter. "Cooling Film Tempers Tiny Hot Spots," *Science News*, November 3, 2001.

Wharton, Ken (Lawrence Livermore National Laboratory). Personal communications, November 2001.

Author's note: Following are the calculations used to determine the size requirements of quantum dots:

Energy (meV) = Temperature (K) * 86 (meV / K)

Area (nm^2) = 7e6 (nm^2* meV) / Energy (meV)

Diameter = 2 * (Area (nm^2) / p)$^{0.5}$

These calculations pertain to a stable quantum dot at room temperature:

Energy = 273 K * 86 meV / K = 23,478 meV

Area = 7e6 / 23,478 meV = 298 nm^2

Diameter = 2 * (298 / p)$^{0.5}$ = 19.48 nm

These pertain to a 2 eV quantum dot:

Area = 7e6 / 2e6 meV = 3.5 nm^2

Diameter = 2 * (3.5 / p)$^{0.5}$ = 2.12 nm

CHAPTER 5

Ashoori, Raymond C. (MIT), Personal communications, June 2001–August 2002.

______. Personal interview, July 13, 2001.

Browne, Malcom W. "Buckyball May Block AIDS Step," *New York Times*, August 3, 1993.

Brucker-Cohen, Jonah. "Liquid Audio," *Wired*, July 2001.

Davidson, Howard L. (Sun Microsystems). Personal interview, June 15, 2001.

Davis, Raymond E., Kenneth G. Gailey, and Kenneth W. Whitten. "The Electronic Structures of Atoms," *Principles of Chemistry* (Saunders College Publishing, 1984).

Encyclopedia Britannica, 2001 Edition: "Magnetism," "Paramagnetism," "Ferromagnetism," "Meissner Effect," "Schrödinger," "Mendeleev," "Transuranium Element," "Rheological Properties."

Freitas, Robert A., Jr. "Molecular Transport and Sortation," "Manipulation and Locomotion," and Appendix A, *Nanomedicine, Vol. 1: Basic Capabilities* (Landes Bioscience, 1999).

Landis, Geoffrey A. Personal communication, September 4, 2001.

Manthey, David. Orbital Viewer for Windows, Orbital Central, 2000.

Marcus, Charles (Harvard). Personal interviews, June 6 and July 16, 2001.

McEuen, Paul L. "Artificial Atoms: New Boxes for Electrons," *Science,* December 5, 1997.

Merkle, Ralph C. "Casing an Assembler." Sixth Foresight Conference on Molecular Nanotechnology, November 10, 1998.

Schedelbeck, Gert, Werner Wegscheider, Max Bichler, and Gerhard Abstreiter. "Coupled Quantum Dots Fabricated by Cleaved Edge Overgrowth: From Artificial Atoms to Molecules," *Science*, December 5, 1997.

Stix, Gary. "Little Big Science," *Scientific American*, September 2001.

Sumitomo Special Metals corporate home page, *http://www.ssmc.co.jp/english/neomax_e.html* (October 2001).

Szabo, Nick. "Re: Artificial Atoms," *sci.nanotech*, December 1, 1993.

Topinka, M. A., B. J. LeRoy, R. M. Westervelt, S. E. J. Shaw, R. Fleichmann, E. J. Heller, K. D. Marankowski, and A. C. Gossard. "Coherent Branched Flow in a Two-Dimensional Electron Gas," *Nature*, March 8, 2001.

University of Sheffield. "Magneto-Rheological Suspensions and Associated Technology." *Proceedings of the 5th International Conference on Electro-Rheological Fluids* (World Scientific Publications, 2000).

CHAPTER 6

Goldhaber-Gordon, David, Michael S. Montemerlo, J. Christopher Love, Gregory J. Opticeck, and James C. Ellenbogen. "Overview of Nanoelectronic Devices," *Proc. IEEE*, Vol. 85, April 1997.

Kleiner, Kurt. "Mutant Viruses Order Quantum Dots," *New Scientist*, May 6, 2002.

McCarthy, Wil. "Once Upon a Matter Crushed," *Science Fiction Age*, May 1999.

______. *The Collapsium* (Del Rey Books/Random House Inc., August 2000).

______. "Programmable Matter," *Nature*, October 5, 2000.

______. "Ultimate Alchemy," *Wired*, October 2001.

______. "Beyond the Periodic Table," *Analog*, January 2002.

Roukes, Michael. "Plenty of Room, Indeed," *Scientific American*, September 2001.

Stix, Gary. "Little Big Science," *Scientific American*, September 2001.

Szabo, Nick. "Re: Artificial Atoms," *Usenet sci.nanotech*, December 1, 1993.

Weiss, Peter. "Circuitry in a Nanowire," *Science News*, February 9, 2002.

______. "Shrinking Toward the Ultimate Transistor," *Science News*, August 10, 2002.

Whitesides, George M., and J. Christopher Love. "The Art of Building Small," *Scientific American*, September 2001.

CHAPTER 7

Bloomfield, Louis A. *How Things Work: The Physics of Everyday Life* (John Wiley & Sons, July 1996).

Canadian Forces Nuclear, Biological and Chemical Defence Equipment Manual, "Individual Protection Ensemble," B-GG–005–004/AF–011 (Book 1 of 2).

Encyclopedia Britannica, 2001 Edition: "Giza, Pyramids of," "Architecture, Techniques," "History of Cement," "Luray Caverns."

Glover, Thomas J. *Pocket Ref* (Sequoia Publishing Inc., 1989).

NASA Surface Meteorology and Solar Energy Data Set, *http://eosweb.larc.nasa.gov/sse/* (2001).

National Science Foundation. "Microscopic Magnetorheological Chains Able to Save Tall Buildings," *NSF Engineering Online News*, April 2001.

Turtle Homes corporate web page, *www.turtlehomes.com* (November 2001).

United States Department of Energy. "Changes in Energy Usage in Residential Housing Units," *http://www.eia.doe.gov* (2001).

Wolfe, William L., and George J. Zissis. *The Infrared Handbook*, rev. ed. Office of Naval Research, Department of the Navy (United States), 1985.

CHAPTER 8

Anderson, Mark K. "Quantum Mechanics' New Horizons," *Wired News*, July 2, 2001.

Austin, Sam M., and George F. Bertsch. "Halo Nuclei," *Scientific American*, June 1995.

"Cloaking: Now You See Me, Now You Don't," *Viewzone.com* (November 1998).

de Garis, Hugo, and J. Storrs Hall. "Picotech? Femtotech?" *sci.nanotech* (December 2, 1993).

Drexler, K. Eric. *Unbounding the Future* (William Morrow and Company, 1991).

Edwards, Kenneth M. (AFRL). Personal interview, September 16, 2001.

Egan, Greg. *Diaspora* (Harper Prism Books, February 1998).

______. Personal communication, July 1, 2001.

Encyclopedia Britannica, 2001 Edition: "Diamagnetism," "Textile: Creping," "Holography," "Catalyst," "Enzyme," "Camouflage."

Frain, John. "Vibration Control Using CMAC Neural Networks," Master's Project Electrical Engineering, April 20, 1999.

Freitas, Robert A., Jr. "Manipulation and Locomotion," *Nanomedicine, Vol. I: Basic Capabilities* (Landes Bioscience, 1999).

Geim, Andrey. "Everyone's Magnetism," *Physics Today*, September 1998.

Geim, A. K., M. D. Simon, M. I. Boamfa, and L. O. Heflinger. "Magnet Levitation at Your Fingertips," *Nature*, July 22, 1999.

Hall, J. Storrs. "Utility Fog: A Universal Physical Substance," *Vision–21: Interdisciplinary Science and Engineering in the Era of Cyberspace*, NASA-CP–10129, pp. 115–126 (1993).

Hameroff, Stuart, Sean Rasmussen, and Bengt Mansson. "Molecular Automata in Microtubules: Basic Computational Logic of the Living State?"

Artificial Life, ed. Christopher G. Langdon (Addison-Wesley Publishing Company, September 1987).

NASA Jet Propulsion Laboratory. "Adaptive Camouflage," *NASA Tech Briefs*, NPO–20706, August 2000.

Pan, Jinato. "MEMS and Reliability," *18–849b Dependable Embedded Systems*, Carnegie Mellon University, Spring 1999.

Penrose, Roger. *The Emperor's New Mind* (Viking Penguin, December 1990).

Reuters, Inc. "Heeding Your Wardrobe," December 8, 2000.

Ruckman, Chris. "Active Noise Control FAQ," Usenet newsgroup, *alt.sci.physics.acoustics*, March 14, 1996.

Schowengerdt, Richard Neal. "Cloaking Using Electro-Optical Camouflage," Patent No. 5,307,162, issued April 26, 1994.

Seng, Yvonne. "Geek Chic: The Dawning of Wearable Tech," Digital Living Today, November 2001.

Simon, M. D., and A. K. Geim. "Diamagnetic Levitation: Flying Frogs and Floating Magnets," *Journal of Applied Physics,* Vol. 87, 6200 (2000).

Smith, Gerald (Positronics Research). Personal interview, November 6, 2001.

Stix, Gary. "Little Big Science," *Scientific American*, September 2001.

AFTERWORD

Banis, Bud. "The Economics of Publishing Your Book" (Science Humanities Press, 1999).

Brown, Charles. "Vernor Vinge: Singular Progress," *Locus*, January 2001

Emanuelson, Jerry. *The Life Extension Manual* (Colorado Futurescience, 2001).

Encyclopedia Britannica, 2001 Edition: "Encyclopedia," "Prometheus."

Library of Congress home page, *www.loc.gov* (December 2001).

Vinge, Vernor. *Marooned in Realtime* (Baen Books, 1986).

APPENDIX B: PATENT APPLICATION

IN THE COURSE OF DEVELOPING the ideas in this book, Dr. Gary Snyder and I eventually determined that our "wellstone" was a patentable—though not immediately producible—invention. Since a working prototype could not be demonstrated, we submitted a document known as a Provisional Patent Application (PPA) to the United States Patent and Trademark Office in mid-2001. For legal purposes, this is considered a "reduction to practice" and is equivalent to a prototype so long as a Regular Patent Application (RPA) is filed within one year.

Our RPA evolved alongside the early drafts of this book, and some information contained in the RPA was too technical, too obscure, too general, too legalistic, or otherwise not immediately relevant to the text of the book. Nevertheless, for readers interested in quantum dots and their applications, or simply interested in the patent application process itself, the document may be somewhat interesting in its own right. What follows is the RPA as originally submitted, format errors and all. Some material has since been expanded, deleted, or reworded at the request of the Patent and Trademark Office, but for my taste the original document is more interesting and informative.

Any inquiries should be directed to Galileo Shipyards, LLC, c/o Brad J. Hattenbach, Esq., Heimbecher & Associates, Union Plaza, Suite 316, 200 Union Boulevard, Lakewood, CO 80228. (*brad@heimbecher.com*).

PATENT APPLICATION OF
Wil McCarthy and Gary E. Snyder
FOR
QUANTUM DOT FIBER

CROSS-REFERENCE TO RELATED APPLICATIONS

This application is entitled to the benefit of Provisional Patent Application Ser.#60/312264, filed 13 August 2001.

BACKGROUND—FIELD OF INVENTION

This invention relates to a fiber whose exterior surface has quantum dots attached to it. The invention has particular but not exclusive application in materials science, as a programmable dopant which can be placed inside bulk materials and controlled by external signals.

BACKGROUND—DEFINITIONS AND THEORY OF OPERATION

The fabrication of very small structures to exploit the quantum mechanical behavior of charge carriers (e.g., electrons or electron "holes") is well established. Quantum confinement of a carrier can be accomplished by a structure whose linear dimension is less than the quantum mechanical wavelength of the carrier. Confinement in a single dimension produces a "quantum well," and confinement in two dimensions produces a "quantum wire."

A quantum dot (QD) is a structure capable of confining carriers in all three dimensions. Quantum dots can be formed as particles, with a dimension in all three directions of less than the de Broglie wavelength of a charge carrier. Such particles may be composed of semiconductor materials (including Si, GaAs, InGaAs, InAlAs, InAs, and other materials), or of metals, and may or may not possess an insulative coating. Such particles are referred to in this document as "quantum dot particles." A quantum dot can also be formed inside a semiconductor substrate, through electrostatic confinement of the charge carriers. This is accomplished through the use of microelectronic devices of various design (e.g., a nearly enclosed gate electrode formed on top of a P-N-P junction). Here, the term "micro" means "very small" and usually expresses a dimension less than the order of microns (thousandths of a millimeter). The term "quantum dot device" refers to any apparatus capable of generating a quantum dot in this manner. The generic term "quantum dot," abbreviated QD in certain drawings, refers to any quantum dot particle or quantum dot device.

Quantum dots can have a greatly modified electronic structure from the corresponding bulk material, and can serve as dopants. Because of their unique properties, quantum dots are used in a variety of electronic, optical, and electro-optical devices.

Kastner (1993) points out that the quantum dot can be thought of as an "artificial atom," since the carriers confined in it behave similarly in many ways to electrons confined by an atomic nucleus. The term "artificial atom" is now in common use, and is often used interchangeably with "quantum dot." However, for the purposes of this document, "artificial atom" refers specifically to the pattern of confined carriers (e.g., an electron gas), and not to the particle or device in which the carriers are confined.

The term "quantum dot fiber" refers to a wire or fiber with quantum dots attached to, embedded in, or forming its outer surface. This should not be confused with a quantum wire, which is a structure for carrier confinement in two dimensions only.

BACKGROUND—DESCRIPTION OF PRIOR ART

Quantum dots are currently used as near-monochromatic fluorescent light sources, laser light sources, light detectors (including infra-red detectors), and highly miniaturized transistors, including single-electron transistors. They can also serve as a useful laboratory for exploring the quantum mechanical behavior of confined carriers. Many researchers are exploring the use of quantum dots in artificial materials, and as programmable dopants to affect the optical and electrical properties of semiconductor materials.

Kastner (1993) describes the future potential for "artificial molecules" and "artificial solids" composed of quantum dot particles. Specifics on the design and functioning of these molecules and solids are not provided. Leatherdale et al. (2000) describe, in detail, the fabrication of "two- and three-dimensional. . .artificial solids with potentially tunable optical and electrical properties." These solids are composed of colloidal semiconductor nanocrystals deposited on a semiconductor substrate. The result is an ordered, glassy film composed of quantum dot particles, which can be optically stimulated by external light sources, or electrically stimulated by attached electrodes, to alter its optical and electrical properties. However, these films are extremely fragile, and are "three dimensional" only in the sense that they have been made up to several microns thick. In addition, the only parameter which can be adjusted electrically is the average number of electrons in the quantum dots. Slight variations in the size and composition of the quantum dot particles mean that the number of electrons will vary slightly between dots. However, on average the quantum dot particles will all behave similarly.

The embedding of metal and semiconductor nanoparticles inside bulk materials (e.g., the lead particles in leaded crystal) is also well established. These nanoparticles are quantum dots whose characteristics are determined by their size and composition, and they serve as dopants for the material in which they are embedded, to alter selected optical or electrical properties. However, there is no means or pathway by which these quantum dot particles can be stimulated electrically. Thus, the doping characteristics of the quantum dots are fixed at the time of manufacture.

However, the prior art almost completely overlooks the broader materials-science implications of quantum dots. The ability to place programmable dopants in a variety of materials implies a useful control over the bulk properties of these materials. This control could take place not only at the time of fabrication, but also at the time of use, in response to changing needs and conditions. However, there is virtually no prior art discussing the use, placement, or control of quantum dots in the interior of bulk materials. Similarly, there is no prior art discussing the placement of quantum dots on the outside of an electrically or optically conductive fiber. There are hints of these concepts in a handful of references, discussed below:

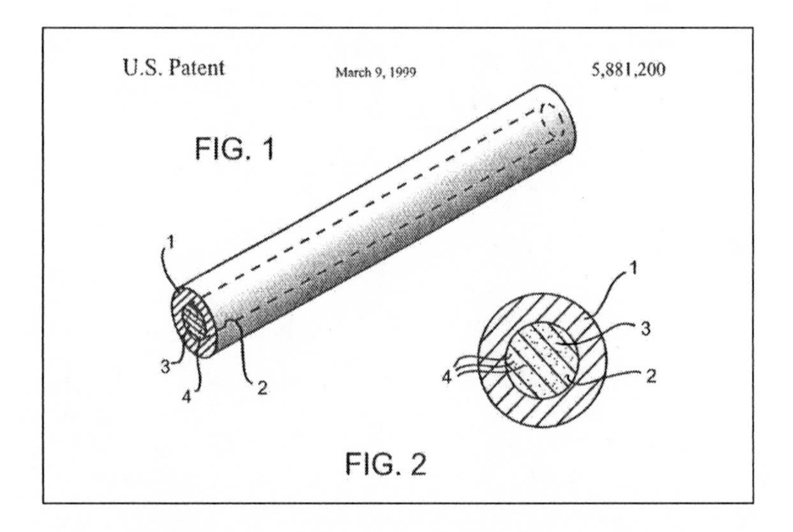

U.S. patent 5,881,200 to Burt (1999) discloses an optical fiber (1) containing a central opening (2) filled with a colloidal solution (3) of quantum dots (4) in a support medium. The purpose of the quantum dots is to produce light when optically stimulated, for example, to produce optical amplification or laser radiation. The quantum dots take the place of erbium atoms, which can produce optical amplifiers when used as dopants in an optical fiber. This fiber could be embedded inside bulk materials, but could not alter their properties since the quantum-dot dopants are enclosed inside the fiber. In addition, no means is described for exciting the quantum dots electrically. Thus the characteristics of the quantum dots are not programmable, except in the sense that their size and composition can be selected at the time of manufacture.

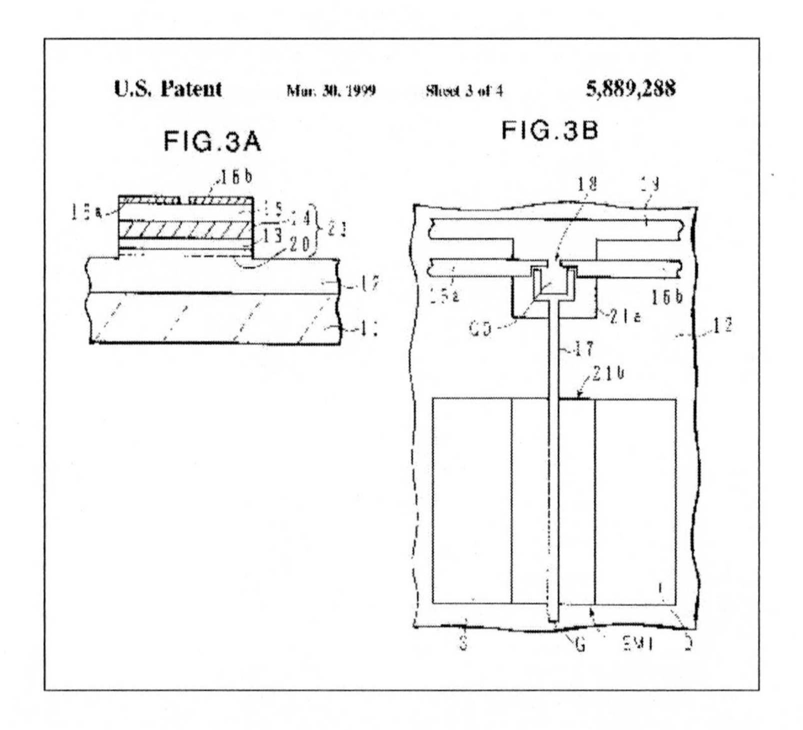

U.S. Patent 5,889,288 to Futasugi (1999) discloses a semiconductor quantum dot device which uses electrostatic repulsion to confine electrons. This device consists of electrodes (16a, 16b, and 17) controlled by a field effect transistor, both formed on the surface of a quantum well on a semi-insulating substrate (11). This arrangement permits the exact number of electrons trapped in the quantum dot (QD) to be controlled, simply by varying the voltage on the gate electrode (G). This is useful, in that it allows the "artificial atom" contained in the quantum dot to take on characteristics similar to any natural atom on the periodic table, including transuranic and asymmetric atoms which cannot easily be created by other means. Unfortunately, the two-dimensional nature of the electrodes means that the quantum dot can exist only at or near the surface of the wafer, and cannot serve as a dopant to affect the wafer's interior properties.

Turton (1995) describes the possibility of placing such quantum dot devices in two-dimensional arrays on a semiconductor microchip. This practice has since become routine, although the spacing of the quantum dot devices is typically large enough that the artificial atoms formed on the chip do not interact significantly. Such a chip also suffers from the limitation cited in the previous paragraph: its two-dimensional structure prevents its being used as a dopant except near the surface of a material or material layer.

Goldhaber-Gordon et. al (1997) describe what may be the smallest possible single-electron transistor. This consists of a "wire" made of conductive C_6 (benzene) molecules, with a "resonant tunneling device" inline which consists of a benzene molecule surrounded by CH_2 molecules which serve as insulators. The device is described (incorrectly, I believe) as a quantum well rather than a quantum dot, and is intended as a switching device (transistor) rather than a confinement mechanism for charge carriers. However, in principle the device should be capable of containing a small number of excess electrons and thus forming a primitive sort of artificial atom.

McCarthy (1999), in a science fiction story, includes a fanciful description of "wellstone," a form of "programmable matter" made from "a diffuse lattice of crystalline silicon, superfine threads much finer than a human hair," which use "a careful balancing of electrical charges" to confine electrons in free space, adjacent to the threads. This is probably physically impossible, as it would appear to violate Coulomb's Law, although I do not wish to be bound by this. Similar text by the same author—myself—appears in McCarthy (August 2000) and McCarthy (05 October 2000). Detailed information about the composition, construction, or functioning of these devices is not given.

The first detailed and technically rigorous discussion of a quantum dot fiber occurs in Provisional Patent Application Ser.#60/312264, filed 13

August 2001 by myself. The Provisional Patent Application forms the basis of this patent application.

SUMMARY

In accordance with the present invention, a quantum dot fiber comprises a fiber containing one or more control wires, which control quantum dots on the exterior surface of the fiber.

OBJECTS AND ADVANTAGES

Accordingly, several objects and advantages of the present invention are:

(a) that it provides a three-dimensional structure for quantum dots which can be considerably more robust than a nanoparticle film. For example, a contiguous glass fiber or metal wire is held together by atomic bonds, as opposed to the much weaker Van der Waals forces which hold nanoparticle films together.
(b) that it provides a method for the electrical and/or optical stimulation of quantum dot particles embedded inside bulk materials. The fiber can consist of, or include, one or more metal wires or optical conduits which are electrically and/or optically isolated from the material in which they are embedded. These pathways can branch directly to the surfaces or interiors of the quantum dot particles, providing the means to stimulate them.
(c) that is provides a method for embedding and controlling electrostatic quantum dot devices (and potentially other types of quantum dot devices) inside bulk materials, rather than at their surfaces.
(d) that it permits the doping characteristics of quantum dots inside a material to be controlled by external signals, and thus varied by a user at the time of use. Thus, the properties of the bulk material can be tuned in real time, in response to changing needs or circumstances.
(e) that the quantum dot fiber can be used outside of bulk materials, in applications where quantum dots, quantum wires, and nanoparticle films

are presently used. For example, the quantum dot fiber can serve as a microscopic light source or laser light source which is both long and flexible.

(f) that multiple quantum dot fibers can be arranged on a surface to produce two-dimensional materials analogous to nanoparticle films, but much stronger.

(g) that multiple quantum dot fibers can be woven, braided, or otherwise arranged into three-dimensional structures whose properties can be adjusted through external signals, forming a type of "programmable matter" which is a bulk solid with electrical and optical properties (and potentially other properties such as magnetic, mechanical, and chemical properties) that can be tuned in real time through the adjustment of artificial atoms.

(h) that the resulting programmable materials, unlike nanoparticle films, can contain artificial atoms of numerous and wildly different types, if desired. Thus, the number of potential uses for the quantum dot fiber materials is vastly greater than for the materials based on nanoparticle films.

DRAWING FIGURES

In the drawings, closely related figures have the same number but different alphabetic suffixes, except for figures 1 and 2 from the prior art, which are closely related.

Figs 1 and 2 are from the prior art, U.S. patent 5,881,200 to Burt (1999), showing an optical fiber containing a central opening filled with a colloidal solution of quantum dots in a support medium.

Figs 3a and 3b are from the prior art, U.S. patent 5,889,288 to Futasugi (1999), showing a semiconductor quantum dot device which uses electrostatic repulsion to confine electrons.

Figs 4a and 4b are from the present invention, in its preferred embodiment. This is a multilayered microscopic fiber which includes a quantum well, surface electrodes which form quantum dot devices, and control wires to carry electrical signals to the electrodes.

Figs 5a and 5b disclose an additional embodiment of the present invention, in which the quantum dot devices (quantum well and electrodes) on the fiber's surface are replaced with quantum dot particles.

Figs 6a and 6b disclose a variant of this embodiment, in which the fiber comprises a single control wire with quantum dot particles attached to its exterior surface.

Figs 7a and 7b disclose still another alternative embodiment of the present invention, comprising an ordered chain of quantum dot particles alternating with control wire segments.

REFERENCE NUMERALS IN DRAWINGS

Reference numerals for the prior art are not included here. The reference numerals for the present invention are as follows:

(30) Surface electrodes
(31) Positive layers of quantum well
(32) Negative layer of quantum well
(33) Memory layer, comprising microscopic transistors to switch electrodes on and off. This layer is optional, since this switching can be accomplished external to the fiber.
(34) Control wires
(35) Insulator
(36) Control wire branches to fiber surface
(37) Quantum dot particles
(38) Control wire segments
(QD) Quantum dot region

Please note that Figures 1–3 are from the prior art, and are included for reference in the Prior Art section of this specification. To prevent confusion, the figures for the present invention, in the drawing pages below, are numbered 4 and above.

DESCRIPTION—FIGS. 4A AND 4B—PREFERRED EMBODIMENT

Figures 4a (isometric view) and 4b (end view) show a preferred embodiment of the invention, which is a fiber containing control wires (34) in an insulating medium (35), surrounded by layers of semiconductor or other materials (31) and (32) which form a quantum well, plus an optional memory layer (33). The central layer (32) of the quantum well must be smaller in thickness than the de Broglie wavelength of the charge carriers to be confined in it. For an electron at room temperature, this would be approximately 20 nanometers. Thicker quantum wells are possible, although they will only operate at temperatures colder than room temperature. Thinner quantum wells will operate at room temperature, and at higher temperatures so long as the de Broglie wavelength of the carriers does not exceed the thickness of the confinement layer (32).

The surface of the fiber includes conductors which serve as the electrodes (30) of a quantum dot device, which confine charge carriers in the quantum well into a small space or quantum dot (QD), forming an artificial atom. The electrodes (30) are powered by control wire branches (36) reaching from the control wires (34) to the fiber's surface. The control wires and control wire branches would normally be electrical conductors, although in principle they could be made of other materials, such as semiconductors or superconductors.

The memory layer (33) comprises microscopic transistors which serve as switches, and which are capable of turning voltages to the surface electrodes (30) on and off. This layer is optional, since this switching can be accomplished external to the fiber. However, it is included here for clarity.

Note that the exact arrangement of the various layers can be slightly different than is depicted here, without altering the essential functioning of the quantum dot fiber. For example, the cross-section may be any oval or polygon shape, and the insulated control wires need not be located at the fiber's center, although that seems to be the most convenient place to locate them.

FIGS 5A–7B—ADDITIONAL EMBODIMENTS

Figures 5a (isometric view) and 5b (end view) show an additional embodiment, in which control wire segments (38) alternate with quantum dot particles (37). The dimensions of both the wire segments and the quantum dot particles, while generally microscopic, could cover a broad range of values while retaining useful properties for the quantum dot fiber.

Figures 6a (isometric view) and 6b (end view) show another additional embodiment, in which quantum dot particles (37) are attached to the surface of a non-insulated control wire (34). In general, this wire would be an electrical conductor, semiconductor, or superconductor, but could in principle be another type of conduit for carrying energy to stimulate the quantum dot particles. Dimensions can once again cover a broad range of microscopic values.

Figures 7a (isometric view) and 6b (end view) show still another additional embodiment, in which the fiber comprises multiple control wires (34) surrounded by insulation (35), with control wire branches (36) leading to quantum dot particles (37) on the surface of the fiber. For clarity, an optional memory layer (33) is included as well. In this embodiment, the control wires could be conductors, semiconductors, or superconductors, but could also be optical fibers, or other types of conduits for carrying energy to stimulate the quantum dot particles (37). Again, the dimensions can cover a broad range of microscopic values while retaining useful optical, electrical, and other properties for the quantum dot fiber.

ALTERNATIVE EMBODIMENTS

There are various possibilities for making the quantum dot fiber of different materials, and in different configurations. The most advantageous configurations are the smallest, since smaller quantum dots can contain charge carriers at higher energies and thus display atom-like behavior at higher temperatures. The smallest conceivable quantum dot fiber would be similar in design to the single-electron transistor described in Goldhaber-Gordon et. al (1997), although molecules the size of benzene rings or smaller, if employed as quantum dot particles, will be unable to hold large numbers of excess charge carriers. This limits their usefulness in generating artificial atoms. A somewhat larger but more practical design is to employ electrically conductive nanotubes, such as a carbon nanotubes, as the control wire segments (38), and fullerene-type molecules, such as carbon fullerenes, as the quantum dot particles (37).

ADVANTAGES

From the description above, our quantum dot fiber can be seen to provide a number of cabailities which are not possible with the prior art:

(a) The ability to place programmable dopants in the interior of bulk materials.
(b) The ability to control the properties of these dopants in real time, through external signals. In contrast, the properties of dopants based solely on quantum dot particles can only be controlled at the time of manufacture.
(c) The ability to form programmable materials containing "artificial atoms" of diverse types. In contrast, programmable materials based on nanoparticle films can contain only multiple instances of one "artificial element" at a time.

Also from the above description, several advantages over the prior art become evident:

(d) Materials based on quantum dot fibers will, in general, be much stronger than materials based on nanoparticle films.

(e) Quantum dot fibers can be used in numerous applications where quantum dots and quantum wires are presently employed. However, the quantum dot fiber provides isolated energy channels for the optical or electrical stimulation of the quantum dots, permitting the dots to be excited without also affecting the surrounding medium or materials. For example, light can be passed through a quantum dot by means of the fiber, without also being shined on or through surrounding areas. Similarly, an electrical voltage can be channeled to a quantum dot without passing through the surrounding medium. Thus, quantum dot fibers can be used in numerous applications where quantum dot devices or particles would prove disruptive.

OPERATION—FIGS 4A AND 4B

The preferred manner of using the quantum dot fiber is to place a fiber or a plurality of fibers, as needed, inside a bulk material (e.g., a semiconductor), or to weave or braid them together into a two- or three-dimensional structure. Material layers (31) and (32) form a quantum well, which traps charge carriers in a quantum (wavelike) manner in the central layer (32).

Voltages (or other energy if appropriate) are then passed through the control wires (30) from an external source. These voltages pass from the control wires to the control wire branches (36), where they are carried to electrodes (30) on the surface of the fiber. Alternatively, the control wire branches may pass through an optional memory layer (33) which consists of transistors or other switches which are capable of switching the voltage pathways open or closed. From the memory layer, the control wire branches would then lead to the electrodes at the surface of the fiber. Once the voltage reaches the electrodes, it creates an electrostatic repulsion which affects the carriers trapped in the quantum well, herding them into small areas known as quantum dots, where they form artificial atoms.

Adjustment of the voltages on the electrodes can then affect the characteristics of the artificial atoms, including:

(a) size
(b) shape or symmetry
(c) number of charge carriers
(d) energy levels of the carriers

The resulting changes in the artificial atom can dramatically affect its properties as a dopant.

Depending on the number of control wires inside the fiber and the number of quantum dot devices along its surface, the artificial atoms located near the fiber's surface (in the confinement layer 32) may all be identical, may represent multiple "artificial elements" in regular or irregular sequences, or may all be different.

OPERATION—FIGS 5A AND 5B

The operation of this embodiment is very similar to the previous one, with the exception that the carriers are confined in quantum dot particles (37) rather than by electrostatic repulsion and a quantum well. Voltages (or optical energy or other energy) are passed through the control wires (34) from an external source, and brought to the fiber's surface via control wire branches (36). These voltages are then carried to the quantum dot particles, in order to stimulate them. This stimulation can then affect the properties of the artificial atoms contained in the quantum dot particles, including:

(a) number of carriers
(b) energy levels of the carriers

As before, the resulting changes in the artificial atom can dramatically affect its properties as a dopant.

Depending on the number of control wires inside the fiber and the number of quantum dot particles along its surface, the artificial atoms located in the quantum dot particles may all be identical, may represent multiple "artificial elements" in regular or irregular sequences, or may all be different.

OPERATION—FIGS 6A AND 6B

The operation of this embodiment is similar to the previous one, with the exception that the fiber comprises a single control wire (34), with quantum dot particles (37) attached to its outer surface. The quantum dots are stimulated by voltage (or optical energy or other energy) passing through the control wire. This stimulation can then affect the properties of the artificial atoms contained in the quantum dot particles, including:

(a) number of carriers

(b) energy levels of the carriers

As before, the resulting changes in the artificial atom can dramatically affect its properties as a dopant.

The capabilities of this embodiment are more limited, in that (barring minor variations in the size and composition of the quantum dot particles) all the artificial atoms along the fiber will have the same characteristics. In some cases it may be necessary to place a high impedance in series with the fiber's control wire in order for a voltage to drive charge carriers into the quantum dots.

OPERATION—FIGS 7A AND 7B

The operation of this embodiment is similar to the previous one, with the exception that the quantum dot particles (37) are not attached to the surface of the fiber, but are an integral part of its structure, alternating with control wire segments (38). A voltage (or optical energy or other energy) is passed through the control wire, and passes directly into and through the quantum dot particles, stimulating them. This stimulation can then affect the properties of the artificial atoms contained in the quantum dot particles, including:

(a) number of carriers

(b) energy levels of the carriers

As before, the resulting changes in the artificial atom can dramatically affect its properties as a dopant.

The capabilities of this embodiment are even more limited than the previous one, in that resistive losses across each quantum dot particle will cause the voltage to drop significantly across each segment of the fiber. Thus, each successive artificial atom along the fiber's length will have a lower voltage (or illumination or other excitation) than the one before it. Thus, the artificial atoms cannot be individually controlled and will not be identical. Instead, the user may select a sequence of artificial elements, of successively lower energies, to be presented by the fiber.

CONCLUSION, RAMIFICATIONS, AND SCOPE

Accordingly, the reader will see that the quantum dot fiber of this invention can be used as a programmable dopant inside bulk materials, as a building block for new materials with unique properties, and as a substitute for quantum dots and quantum wires in various applications (e.g., as a light source or laser light source).

Although the description above contains many specificities, these should not be construed as limiting the scope of the invention but merely providing illustrations of some of the presently preferred embodiments of this invention. For example, the fiber could have non-circular shapes in cross-section, including a flat ribbon with quantum dots on one or both sides; the "artificial atoms" could be composed of charge carriers other than electrons; the control wires could be replaced with semiconductor, superconductor, optical fiber, or other conduits for carrying energy; the control wires could be antennas for receiving signals and energy from electromagnetic waves; any of the embodiments listed here could be replicated on a molecular scale through the use of specialized molecules such as carbon nanotube wires and fullerene quantum dot particles; the quantum dots could be other sorts of particles or devices than those discussed herein; the number and relative sizes of the quantum dots with respect to the fiber could be significantly different than is shown in the drawings.

Thus the scope of the invention should be determined by the appended claims and their legal equivalents, rather than by the examples given.

CLAIMS: I claim:

1. In a device for producing quantum effects, comprising:

(a) a material fashioned into an elongated fiber shape, as in a wire, ribbon, or optical fiber

(b) one or more control paths which carry energy along said fiber

(c) quantum dots, whether particles, devices, or other types, on or near the surface of the fiber, which trap and hold a configuration of charge carriers based on the energy or energies in said control paths, thus forming artificial atoms

whereby said fiber can serve as a substitute for quantum dots and quantum wires in existing and future applications, and

whereby the electrical, optical, and possibly other properties such as magnetic, mechanical, and chemical properties of said fiber can be manipulated through adjustment of the energies in the control paths, and

whereby said fiber can be embedded inside a bulk material, to serve as a programmable dopant which is capable of altering the electrical, optical, and possibly other properties of said material in real time based on the energies in said control paths, and

whereby a plurality of said fibers can be woven, braided, or otherwise arranged into two- or three-dimensional structures, creating materials whose characteristics are electrically or optically programmable in real time by means of the energies in said control paths.

2. The device of Claim 1 wherein said control paths are electrical wires, whether conductors, semiconductors, or superconductors, which carry electrical voltages.

3. The device of Claim 1 wherein said control paths are optical fibers carrying light or laser energy.

4. The device of Claim 1 wherein said control paths are radio frequency or microwave antennas.

5. The device of Claim 1 wherein the quantum dots are quantum dot particles

6. The device of Claim 1 wherein the quantum dots are quantum dot devices

7. In a method for controlling dopants in the interior of bulk materials, comprising:

(a) confining charge carriers in a dimension smaller than the de Broglie wavelength of said carriers, such that the carriers assume a quantum wave-like behavior in all three dimensions

(b) carrying electrical or other energy through conduits to said carriers while embedded in a solid material, without said energy directly contacting said material except through said carriers

whereby said carriers form configurations such as artificial atoms, which are capable of serving as programmable dopants to alter the electrical, optical, and possibly other properties such as magnetic, mechanical, and chemical properties, of said material in real time, and

whereby a plurality of said methods can be combined, creating a means for producing materials whose electrical, optical, and possibly other properties such as magnetic, mechanical, and chemical properties can be adjusted in real time.

8. The method of Claim 6 wherein the means of confining said charge carriers is a plurality of quantum dot particles or quantum dot devices, and said conduits are consolidated into fibers to which said quantum dot particles or quantum dot devices are attached.

QUANTUM DOT FIBER

Abstract: A fiber of microscopic diameter, having control wires (34), possibly surrounded by an insulator (35), which pass energy to electrodes (30) on top of material layers (31) and (32) which form quantum dots, or to quantum dot particles (37) on the fiber's surface. The energy passing through the wires stimulates the quantum dots (QD), leading to the formation of "artificial atoms" with real-time tunable properties. These artificial atoms then serve as programmable dopants. The fiber can be used as a programmable dopant inside bulk materials, as a building block for new materials with unique properties, or as a substitute for quantum dots or quantum wires in certain applications.

QUANTUM DOT FIBER

SEQUENCE LISTING

Not applicable.

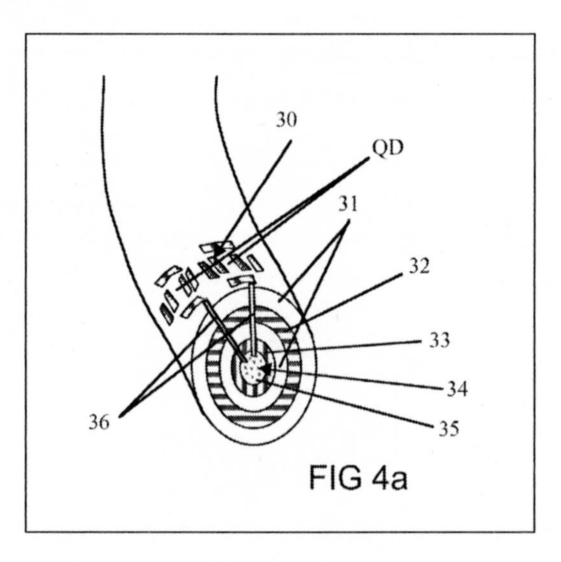

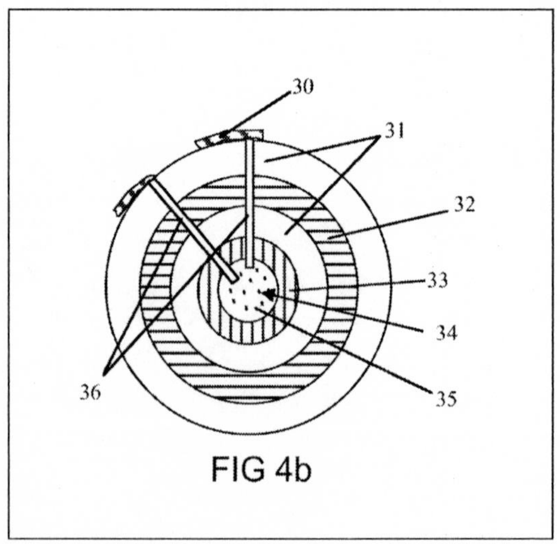

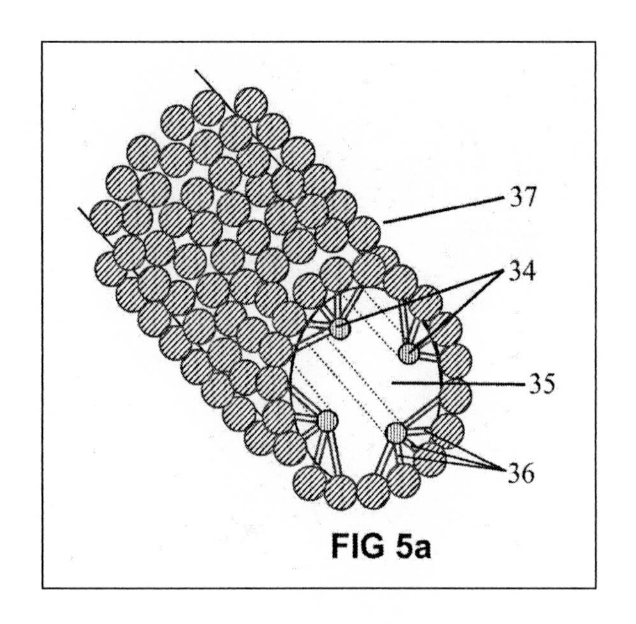

FIG 5a

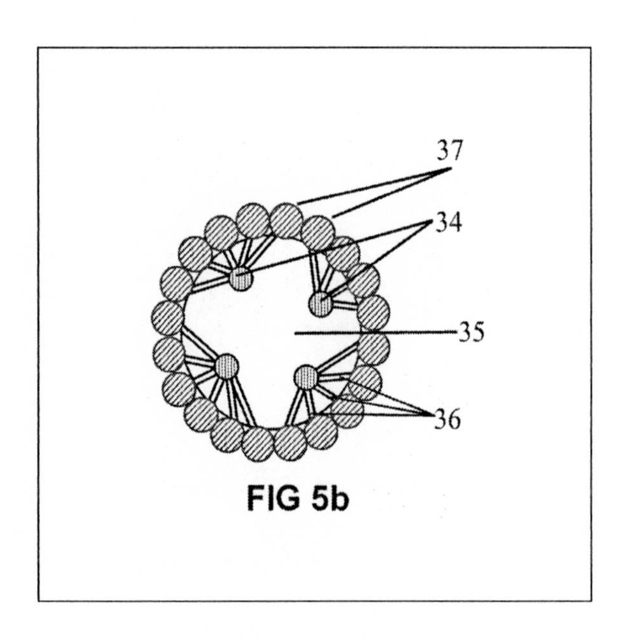

FIG 5b

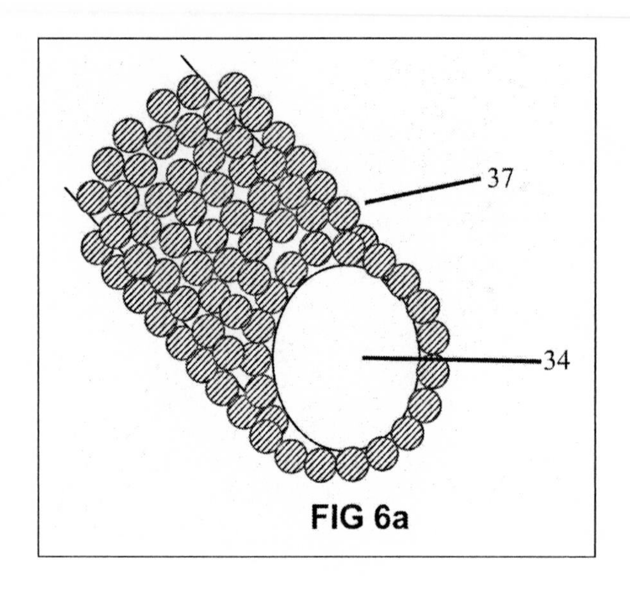

FIG 6a

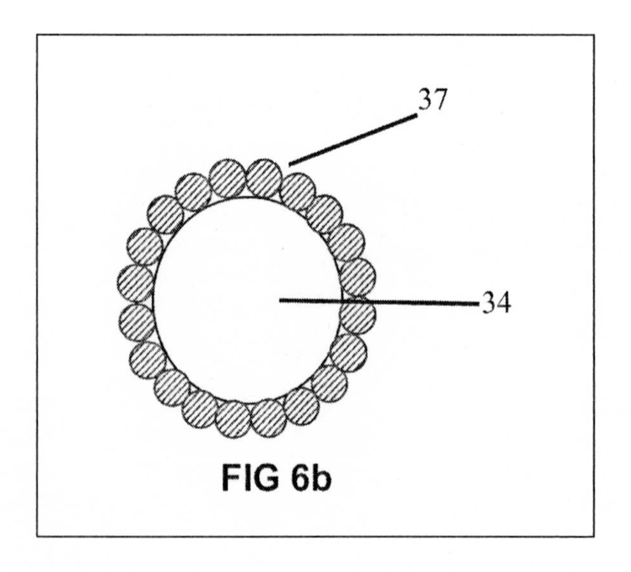

FIG 6b

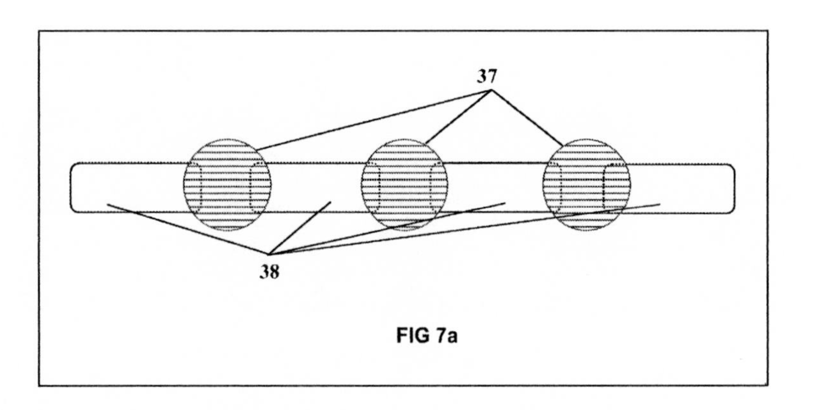

FIG 7a

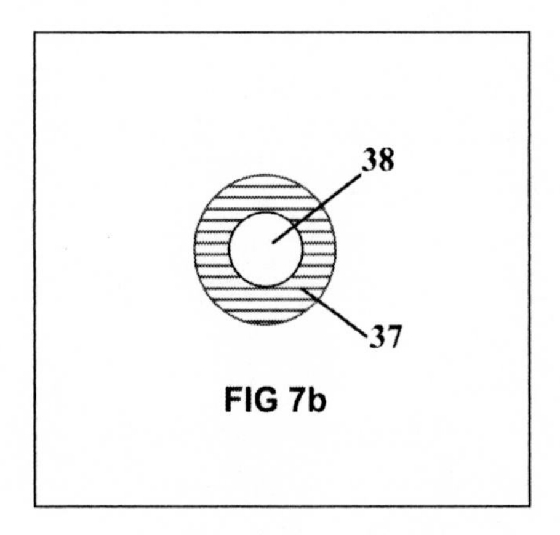

FIG 7b

ACRONYMS

AFM	atomic force microscope
AFRL	Air Force Research Laboratories
BIOSes	Basic Input Output Systems
CCD	charge coupled device
CIMS	Center for Imaging in Mesoscale Structures
DES	Digital Encryption Standard
EM	"Earth Millennium"
ER	electro-rheological
FET	field-effect transistor
GPS	global positioning system
IPO	initial public offering
IR	infrared
LED	light-emitting diode
MEMS	microelectromechanical systems
MOSFETs	metal oxide semiconductor field-effect transistors
MR	magneto-rheological
NIST	National Institute of Standards and Technology
NTT	Nippon Telephone and Telegraph
PPA	Provisional Patent Application
QCA	quantum cellular automaton
QDC	Quantum Del Corporation
RPA	Regular Patent Application
RTG	radioisotope-thermal generator
RTI	Research Triangle Institute
SET	single-electron transistor
UV	ultraviolet
VCs	venture capitalists

INDEX